Lecture Notes in Earth Sciences

ctd. on inside back cover

Lecture Notes in Earth Sciences

Edited by Somdev Bhattacharji, Gerald M. Friedman, Horst J. Neugebauer and Adolf Seilacher

32

Bernd Lehmann

Metallogeny of Tin

Springer-Verlag
Berlin Heidelberg GmbH

Author

Priv.-Doz. Dr. Bernd Lehmann
Freie Universität Berlin, Institut für Angewandte Geologie
Wichernstr. 16, D-1000 Berlin 33

ISBN 978-3-540-52806-7   ISBN 978-3-540-47153-0 (eBook)
DOI 10.1007/978-3-540-47153-0

2132/3140-543210 – Printed on acid-free paper

Mining by fire setting in the Geyer tin deposit, Erzgebirge, 18th century. Sheeted tin-bearing veins with greisenized rims in the Geyer granite stock (Charpentier 1778:Fig. 3). Charpentier is the first to note the importance of hydrothermal overprint in tin ore formation based on the observation of the gradual nature of the rock sequence of vein-greisen-granite

"Das Bekannte überhaupt ist darum,
weil es bekannt ist, nicht erkannt."
(Hegel 1807:25)

Preface

The search for tin dates back to the earliest days of civilization. For about 40 years, world tin mining has oscillated at a level of 150,000-250,000 t Sn/year, with a mine output in 1989 of 210,000 t Sn (MCS 1990). This figure corresponds to a current annual value of about US$ 1.5 billion and places tin ninth on the metal market behind iron, gold, uranium, copper, zinc, silver, platinum and nickel.

Tin deposits belong to the granite-related ore deposit spectrum which includes many metals vital to current and future technologies such as Cu, W, Mo, U, Nb, Ta, Ag, Au, Sb, Bi, As, Pb, Zn, REE, Be, Ga and Li. The granitic rocks associated with tin and tin-tungsten deposits have long been identified as a special group of granites, the so-called tin granites. These rocks provide a unique opportunity to study the magmatic and hydrothermal history of tin ore formation. Tin granites are more easily identifiable as parent rocks for tin (and tungsten) mineralization than is the case for other mineralized granitic rocks such as molybdenum and copper porphyries. The magmatic molybdenum and copper distribution patterns are more complex (control by sulfide solubilities), and commonly obliterated by fluid interaction. The relatively simple situation of tin granites provides therefore an invaluable opportunity to study some metallogenic aspects of magmatic-hydrothermal ore deposits in general.

The present study attempts to develop a general metallogenic model for tin in identifying the essential or relevant processes in tin ore formation. The methodological principle is based on an interplay between a background of some basic petrogenetic concepts and a number of specific local and regional data on tin deposits and tin provinces, with particular reference to those areas with which the author is most familiar with (Bolivia, SE Asia, Europe). This inductive approach condenses the many apparently specific complexities encountered in individual ore deposits to a few major processes of general importance. The inherent reductionism may have a personal bias which is probably inevitable in any simple and broad-scale picture ("Après tout, la raison est bien l'esclave des passions"; Feyerabend 1979:210). The critical problem of the relevance of those factors chosen for our model can be judged by its degree of consistency and predictive capability for new and analogous cases.

This monograph results from a habilitation thesis at Freie Universität Berlin. Continuous encouragement by Hans-Jochen Schneider (Berlin and München) is greatly acknowledged. Field and laboratory work was made possible by Deutscher Akademischer Austauschdienst and technical cooperation projects of Bundesanstalt für Geowissenschaften und Rohstoffe in Hannover. A research grant from Deutsche Forschungsgemeinschaft was instrumental in allowing this text to be prepared. Critical reviews by Philip Candela (College Park, Maryland), John Cobbing (Nottingham), Peter Möller (Berlin), and Helmut Schröcke (München) on parts of the manuscript helped to improve the text. Et surtout merci beaucoup à Evelyne.

Contents

1 Introduction

1.1 General Metallogenic Concepts

Ore deposits are distributed unevenly on Earth and tend to cluster in specific large-scale zones for which de Launay (1913) introduced the term "metallogenic province". The observational background for this situation has been known since the earliest days of mining. Pliny the Elder (A.D. 23-79) states with reference to silver-bearing veins: "...ubicumque una inventa vena est, non procul invenitur alia". (Where a vein is found, there will be another one not far away).

Cotta (1859:236) establishes in one of the first textbooks on ore geology the principle: "Wo einmal eine Erzlagerstätte gefunden wurde, da kann man erwarten, unter analogen Verhältnissen auch mehrere derselben oder ähnlicher Art zu finden, denn die meisten von ihnen pflegen gesellig aufzutreten". (Where an ore deposit has been found, it can be expected to find others of similar type under analogous conditions, because most ore deposits tend to occur gregariously). This simple rule is of fundamental importance in exploration geology, condensed in the familiar saying "If you are hunting elephants, go to elephant country".

A further accentuation of the concept of "metallogenic provinces" is derived from the fact that the province-specific metals are sometimes repeatedly concentrated in the same area by different geological processes and at different times, a situation which has been characterized by Routhier (1967) by the notion of "étagement temporel" (temporal superposition). Examples for such situations are given in Routhier (1980).

The existence of metallogenic provinces, i.e., concentration of ore deposits with similar metal-mineral association on a regional scale, points to the control of ore-forming processes by the regional geological framework. The critical role of such large-scale factors is the subject of metallogenic studies. The broader regional approach offers an important supplement to the more usual detailed investigations of individual ore deposits. The focus on detailed relationships within individual ore deposits has yielded critical data about the ores and their genetic processes. The regional approach addresses the broader relationships that link deposits to one another and provides insights into the regional geological environment which causes locally extreme metal concentrations.

Broad-scale relationships are per se more complex than relationships on a local scale, which makes quantitative modelling difficult. On the other hand, the regional approach makes it possible to filter from the many individual observations those that are specific for the sum of the ore deposits in a given metallogenic province, and that can be regarded as critical factors for ore formation in general.

The observational background of each student of ore deposits is, of course, different. This may introduce a subjective factor into any judgement of a complex sum of individual observations. The acquisition of general principles from a large data set relies on extra-observational background knowledge and the partly intuitive notion of relevance, about which different people may have different ideas. This is the reason for the diversity in metallogenic concepts dealing with identical subjects.

The existence of metallogenic provinces has been interpreted by Schuiling (1967:540) as reflecting a regional geochemical anomaly: "The concept of a metallogenetic province implies the existence of large-scale chemical inhomogeneities in that part of the crust or the mantle from which the ore deposits ultimately were derived".

This statement extends the neutral notion of metallogenic province to the speculative concepts of geochemical province, regional geochemical specialization and metal domain (Routhier 1967). It has been developed by Routhier (1980:46) into the "basic theorem" of metallogeny: "The concentrations of a metal appear at the intersection of a metal domain (actually a volume capable of reaching down to the mantle), bearing during very long periods of time (permanency and heritage) a «metal potential» (that is the primordial metallotect), and of other metallotects, acting as revealers of this potential" (translation in Routhier 1983:42). The term "metallotect" has been defined by Laffitte et al. (1965:3) as "any geological feature or phenomenon associated with lithology, paleogeography, structure, geochemistry, etc. which has contributed to the formation of a mineral concentration" (translation in Routhier 1983:42).

Brimhall and Crerar (1987:235) give a more physicochemically defined concept of ore formation: "...chemical fractionation effects peculiar to fluid/fluid, fluid/rock, and fluid/solute interactions which efficiently extract ore elements from large source regions and quantitatively concentrate them in relatively small physical domains which are preserved for geological periods of time. It is the potential for such repetitive chemical focussing in the vicinity of available sources of thermal energy that drives ore-forming systems to extreme values of reaction progress and fluid dominance. The attainment of end stages of chemical fractionation separates ores from the more common products of petrogenesis".

It is interesting to note that the basic concept of these modern metallogenic theories was put forward a very long time ago. Rößler (1700:5) states in one of the first compilations on ore deposits, referring to vein deposits which at that time were identical to ore deposits in general: "Das Ertz hat seinen Wachsthum aus den Gebürgen und Gestein. Weil aber das Gestein ein festes Corpus, so kan der Gang nicht allenthalben seine Nahrung so vollkömmlich zu sich ziehen, und die Würckung rechte statt finden, wo nicht Flötze, Fälle und Geschicke sich dabey befinden, oder andere Gänge übersetzen, welche alle das Gestein durchschneiden und öffnen". (Condensed translation: Ore has its growth from the rocks. Because rocks are solid, veins develop preferentially in those parts of rocks which have been transformed and opened by faulting).

Zimmermann (1746:105) explains: "Mineralien werden ordentlich im Gesteine erzeuget...Das Gesteine wird nur nach gewissen Strichen und Streifen, welche man Gänge nennet, in Ertzt verwandelt...Dergleichen Gänge, wenn wir sie genau betrachten, nichts anders als ein durchwittertes und mürbe gemachtes Gestein... vorstellet, und wir können sehr wahrscheinlich schliessen, daß dieser Gang vorher ebenfalls mit dem Gestein des gantzen Berges gleiches Wesens gewesen, durch ein auf denen Klüften eindringendes Saltz-Wesen aber also durchdrungen, in seinem Gewebe geändert, aufgeschlossen, und zur künftigen Ertzt-Erzeugung geschickt gemacht worden". (Condensed translation: Ore develops in orderly fashion in the rock. The rock is transformed into ore in certain zones which are called veins. These veins represent brittle and altered rock, and it can be concluded that veins originate from rock which has been transformed by salt fluids percolating on fissures).

Agricola (1546) had already stressed the importance of water in ore formation. His ideas, however, were firmly confined by the theory of the four elements of the Greek philosophers Empedocles and Plato, leaving little room for a chemical understanding of ore deposits.

1.2 The Example Tin

Tin provinces are one of the best examples of metallogenic provinces. They define belts on a 100- to 1000-km scale. Inside tin provinces, the association of tin ore deposits with granitic rocks has long been known (Werner 1791; Zimmermann 1808). Very early studies already emphasize the association of tin and quartz-rich rocks (Rößler 1700), which at that time could not be further detailed (i.e. the non-discrimination of sandstone and granite).

These oldest observations are clouded by theoretical concepts which relate ore formation to astronomy and climate, a result of the then still influential Aristotelian physics. Rößler (1700:19) remarks with respect to tin: "Dieses Metall hat seine Art gerne an kalten Orten" (This metal likes cold places); and Lehmann (1751:12) notes: "Das eintzige Zinn scheinet eine gemäßigte Gegend zu lieben, und es ist daher entweder gar nicht oder wenigstens sehr selten in denen kältern Nord-Ländern zu finden". (Tin appears to love moderate climates and is therefore never or at least rarely found in cold northern countries). Lehmann (1753:203) adds shrewdly: "Zinn ist gerne alleine". (Tin loves to be alone).

Humboldt (1823a) introduces the term "tin granite" as opposed to "normal granite" (in the first edition of his book on global comparative geology, written in French: "granite stannifère" versus "granite primitif").[1]

A first comprehensive treatise on the ore geology of tin and at the same time the first scientific theory of magmatic-hydrothermal ore formation in general is given by Beaumont (1847), helped by the work of Daubrée (1841). Essential findings are:

---

[1] Humboldt fixes the notion of tin granite in a purely chronologic or stratigraphic sense, on the basis of the (erroneous) neptunistic concept that specific metal accumulations and specific rock types characterize specific geological epochs, i.e. different evolutionary stages of sea water (Werner 1791). This concept leads to the systematic theory on time-bound mineral deposits (Karsten 1806). The observation that tin deposits are not only confined to primitive granites ("Ur-Granit" as earliest precipitation from sea water; Goethe 1785) but occur also in the gneiss and mica schist of Gierczyn/Giehren in Silesia (Buch 1802) (see Chap. 5.2), with gneiss at the time generally accepted as the second-oldest rock formed from sea-water precipitation, led Humboldt (1823a,b) to the conclusion that tin-mineralized granites must represent a separate evolutionary stage in between the granite- and gneiss-forming epochs.

1. Tin ore deposits are associated with granites.
2. Tin ore deposits are located preferentially in apical portions of granites and their immediate country rocks. [Daubrée (1841) had already pointed to the fact that tin deposits in Cornwall, in the Bretagne and the Erzgebirge are always confined to a zone ≤500 m from the granite contact].
3. Tin ore deposits are often associated with individual granite bodies pointing out of larger batholiths: "les roches stannifères sont souvent des masses détachées qui ont pointé en dehors des grandes masses granitiques" (Beaumont 1847:1302).
4. The tin ore host rocks are particularly rich in quartz, tourmaline and fluorine-bearing minerals.
5. Tin granites are anomalous in texture and composition: "des monstruités de granite" with a "caractère ultragranitique" (Beaumont 1847:1303).

Daubrée (1841) and Beaumont (1847) define tin granites as granitic rocks in which the metals of the "tin family" are particularly abundant, i.e. Sn, F, B, P, As, W, Mo, Fe. The typical kaolinization in tin granites is interpreted by Daubrée (1841) as a result of hydrothermal alteration by acid fluids. The formation of tin ore deposits is explained by Beaumont (1847) by the action of circulating fluids which leach the chemical components of the ore deposits from the wall rock. The possibility that water and other volatile components may depress the solidus of a granitic melt is discussed, and Beaumont (1847) speculates that tin granites possibly solidify at lower temperatures than normal granites.

The rule of the association of tin deposits with granitic rocks was first derived from European localities only. The validity of this assertion was extended by the statistical analysis of Ferguson and Bateman (1912) on a worldwide basis.

Cotta (1859:680) stressed the important fact that granite magmatism is only exceptionally accompanied by tin mineralization: "...man darf nicht, überall wo granitische Gesteine zu Tage treten, an ihren Grenzregionen auch Zinnerze erwarten, vielmehr ist die Zinnerzbegleitung für die Granite nur eine Ausnahme, während die Granitbegleitung für die Zinnerze eine Regel bildet". (Not everywhere where granitic rocks are found, can tin ore be expected in the contact zones. The occurrence of granitic rocks together with tin ore is an exception, whereas the occurrence of tin ore together with granitic rocks is a rule).

These early observations provide the basis for subsequent investigations which refine the geological-mineralogical-geochemical framework of tin deposits. Sandberger (1885) emphasizes the point that tin granites carry Li-mica (protolithionite and zinnwaldite) with tin contents up to more than 1000 ppm. This aspect of chemical specialization is later developed by Barsukov (1957) on a systematic basis (see below).

A modern and detailed overview of the geology of tin ore deposits is given in Taylor (1979); shorter reviews of general aspects of tin ore formation are provided by Eugster (1985), Kwak (1987), Plimer (1987) and Strong (1988). The more important types of tin ore deposits are compiled in the models of Fig. 1. The ore deposit spectrum can be understood as consequence of a general magmatic-hydrothermal evolution which falls into two broader subjects discussed in detail below:
1. Granitic magmatism of specific petrogenetic type.
2. Associated hydrothermal activity, physically and chemically focussed by the local geological framework and by evolutionary permeability conditions.

All major types of tin ore deposits are associated with highly fractionated granitic rocks in a high-level environment (including the tin pegmatites in Central Africa and SE Asia). The geometric style of the ore deposits is a consequence of local conditions of permeability (tectonic-geological framework plus fluid evolution in the magmatic-hydrothermal system). Examples for these relationships will be given in Chaps. 3 and 4.

1.3 Spatial and Temporal Distribution of Tin Ore Deposits

The province character of tin ore deposits is particularly well developed. Zippe (1857:184) notes: "Das Zinnerz ist nur in sehr wenig Ländern einheimisch". (Tin is indigenous to very few countries). Only four well-defined regions account for 80 % of the cumulative historic tin mine output. These are (Table 1, Fig. 2):

Fig. 1 (next page). Schematic models of the major groups of primary tin ore deposits. After Taylor et al. 1985. 1 tin porphyries (examples: Llallagua and Chorolque, Bolivia; Yinyan, Guangdong, China); 2 skarns and carbonate/sulphide replacements (examples: Cleveland and Renison Bell, Tasmania, Australia; Dachang district, Guangxi, China); 3 veins and sheeted veins (examples: Chojlla, Bolivia; Geevor and Wheal Jane, Cornwall, England; Hermyingyi, Burma); 4 greisens (examples: Altenberg, East Germany; Cinovec, CSSR; Tikus, Indonesia) and pegmatites (examples: Manono, Zaire; Phuket district, Thailand)

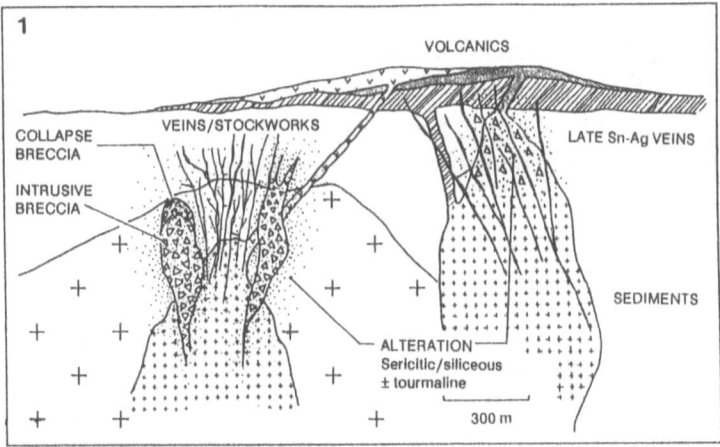

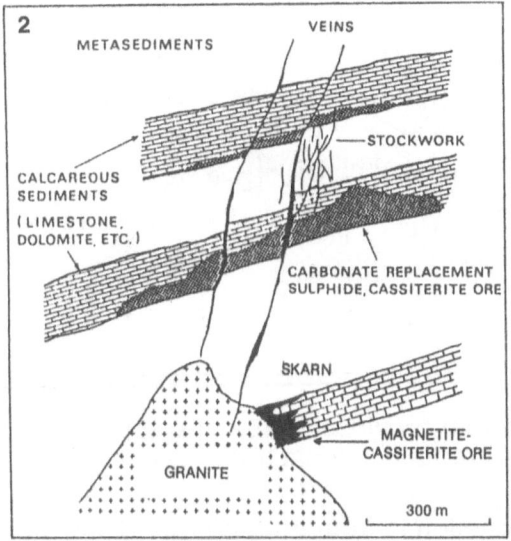

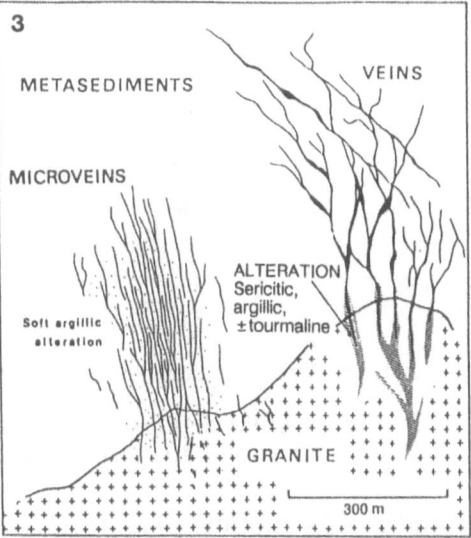

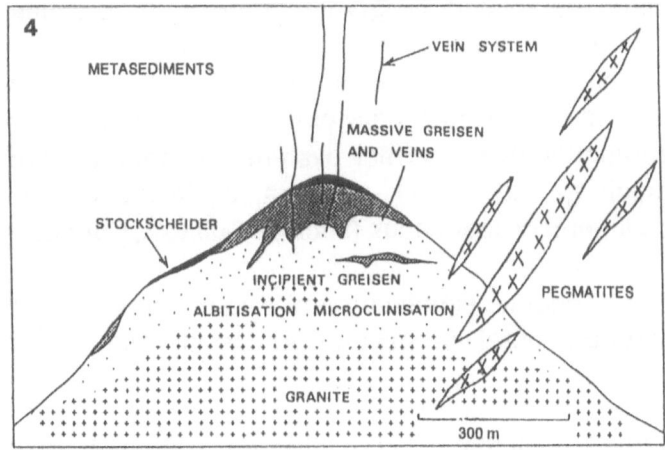

1. The SE Asian tin belt (Burma, Thailand, Malaysia, Indonesia) with a 50 % share of the total world tin production.
2. The Bolivian tin belt (ca. 10 %).
3. The South China tin province (ca. 10 %).
4. The Cornwall tin province (ca. 10 %).

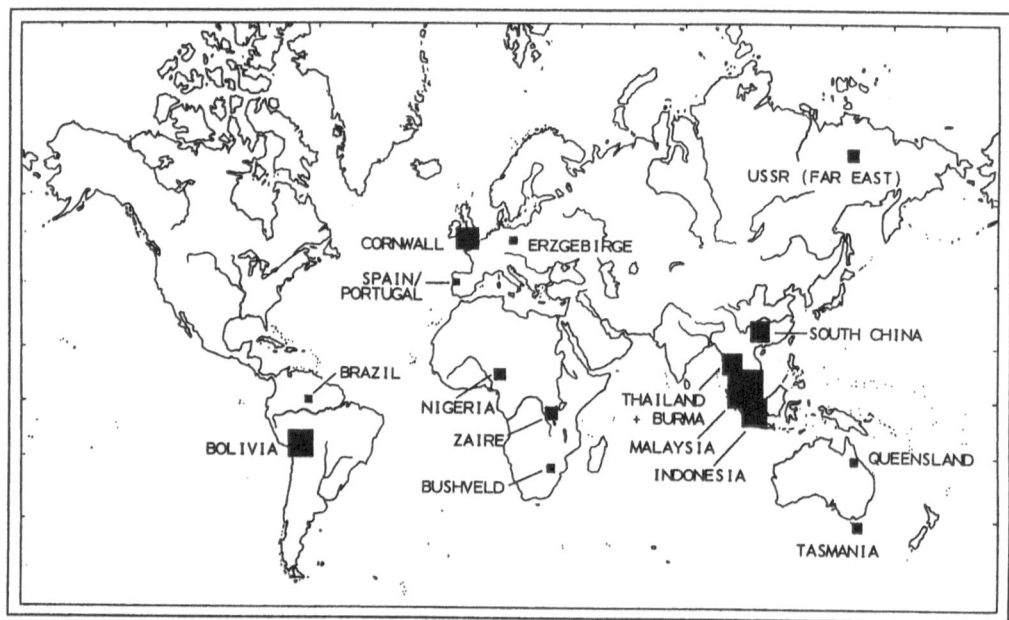

Fig. 2. Geographic distribution of major tin provinces. Area of black squares is proportional to cumulative tin production (data from Table 1)

More than 99 % of the historic tin production is from ore deposits directly or indirectly related to granitic rocks, i.e. granites and their volcanic and subvolcanic equivalents. A small quantity of tin is recovered as a by-product of mining of base-metal massive sulphide deposits. In such deposits, cassiterite is occasionally concentrated at high-temperature vent zones of the submarine-hydrothermal systems of both volcano-sedimentary sequences (such as Kidd Creek, Canada, or Neves Corvo, Portugal) and in clastic sedimentary sequences (such as Sullivan, Canada) (Mulligan 1975).

The economic significance of this mineralization style will increase with the inauguration of the new tin processing plant of the Neves Corvo copper-tin

---

Table 1 (next page). Cumulative tin production up to 1986 (in metric tonnes of metal content) and tin reserve base 1984-1986. Placer deposits are assigned to their primary source deposits

	Location	Production [x 1000 mt]	Reserves [x 1000 mt]
More than 1 Mio t Sn			
1. SE Asian Tin Belt			
200-220 Ma	Malaysia	5200	1200
	Indonesia	2500	1550
	Thailand	300	300
50-100 Ma	Thailand	900	900
	Burma	140	500
2. South China Tin Province			
70-150 Ma	China	1500	1500
3. Bolivian Tin Belt			
200-220 Ma	Bolivia	250	180
12-25 Ma	Bolivia	2000	800
4. Cornwall Tin Province			
270-290 Ma	England	2000	260
0.5-1 Mio t Sn			
5. W Australian Tin Fields			
350-360 Ma	Tasmania	400	230
220-240 Ma	Queensland	200	120
6. Central African Tin Province			
950-1050 Ma	Zaire	450	200
	Rwanda	80	
	Burundi/Uganda/Tanzania	40	
7. Jos Plateau Tin Fields			
540-565 Ma	Nigeria	54	28
160-170 Ma	Nigeria	486	252
0.1-0.5 Mio t Sn			
8. Tin Fields in the Far East of the USSR			
120-170 Ma	USSR	100	300
50-100 Ma	USSR	300	700
9. Erzgebirge Tin Province			
280-305 Ma	Germany/CSSR	300	200
10. Rondônia and Amazonas Tin Provinces			
1000/1500 Ma	Brazil	180	> 1000
11. Iberian Tin Province			
280-300 Ma	Spain/Portugal	150	30
12. Bushveld Tin Province			
1950-2000 Ma	RSA	115	150

Historical tin production data from Reyer (1881), MacAlister (1908), Beyschlag et al. (1910), ITRDC (1938), Ahlfeld (1958), Ahlfeld and Schneider-Scherbina (1964), ITC (1967), Fox (1969), and Metallgesellschaft (1965, 1976, 1989). Reserve data from Crowson (1984, 1986), Thormann and Drew (1988), Seltmann (1990)

mine in mid-1990, which is designed to produce 5000 t/y of tin in concentrate, i.e. 2.5 % of the world tin production. The Neves Corvo massive-sulphide ore deposit is located in the Upper Paleozoic Pyrite Belt of southern Portugal and Spain and has ore reserves of 27 Mio t with 8 wt% Cu, 1 wt% Zn, 0.2 wt% Pb, and 33 ppm Ag, of which 2.8 Mio t contain 2.6 wt% Sn (Carvalho 1986; Tin International 1989; Mining Journal 1990). The geology of this spectacular ore deposit is currently still little known.

The age distribution of tin ore deposits is strongly biased towards the Phanerozoic (Fig. 3). About 90 % of the historic tin production is related to primary tin ore deposits ≤ 300 Ma in age. This is a situation similar to and even more pronounced in molybdenum porphyries, copper porphyries and epithermal Au-Ag ore systems (Meyer 1985).

Such a pattern may result from an accelerating rate of ore formation towards the present, or, alternatively, from continuous reworking/destruction of ore deposits with a constant or variable rate of ore formation. It is evident that the survival rate of an ore deposit in a shallow environment, subjected to uplift and erosion, must be small (fast recycling rate) whereas an ore deposit formed at deep levels of the crust and/or with little uplift and erosion will have a high survival rate. The particularly ephemeral nature of oceanic crust with a half-life of around 60 Ma (including ore deposits within this crust) is shown in Fig. 3.

It is a feature common to both tin ore deposits and molybdenum and copper porphyries, as well as to epithermal ore systems, to be located in tectonic environments very sensitive to erosion. These environments include apical portions of shallow-level intrusions or generally, regions near the Earth's surface, that also experience active uplift in continental margins. The important control of level of erosion is demonstrated by the SE Asian tin province in which primary tin ore deposits are preserved only as relics and are mostly in an early state of erosional dispersion, i.e. in placer deposits. The age distribution of copper and Au-Ag deposits underlines this aspect. These metals are not only related to granitic magmatism in the general porphyry/uplift environment, such as is the case for tin, but are concentrated also in intra- and pericratonic depressions/basins (high survival rate) in which they occur in the Proterozoic (Cu-Pb-Zn-Ag-Au) and Archaean (Au).

The oldest tin ore deposits occur in the Kapvaal craton in southern Africa (Swaziland) in association with Archaean, intracratonic A-type granites (Hunter 1973; Trumbull and Morteani 1986). An intracratonic and anorogenic setting is typical of all major Precambrian tin provinces (Bushveld, Central Africa, Amazonas and Rondônia/Brazil), as opposed to the active margin

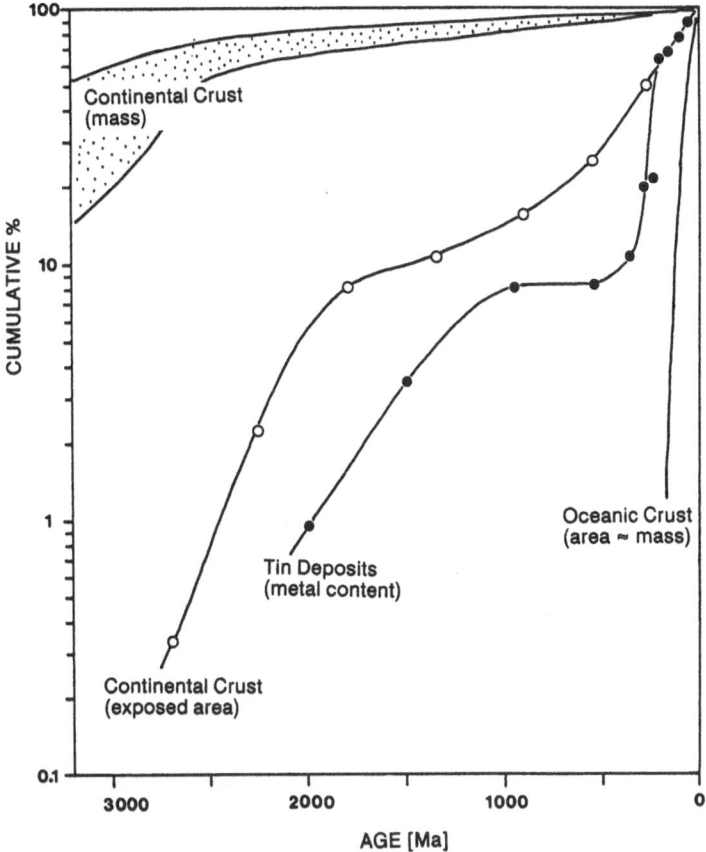

Fig. 3. Age distribution of tin deposits (historic tin production plus reserves; age refers to primary tin accumulation) compared to crustal growth curve (stippled pattern: Taylor and McLennan 1985; Jacobsen 1988) and age distribution of exposed continental crust and oceanic crust. The data for Precambrian crust exposed on land surfaces are from Hurley and Rand (1969), adapted to the global ratio of Phanerozoic/Precambrian exposures of three (Blatt and Jones 1975). Phanerozoic data from Blatt and Jones (1975), data on oceanic crust from Sclater et al. (1981)

setting of most Phanerozoic tin provinces. A systematic treatment of the age relationship of major geotectonic environments has been given by Veizer and Jansen (1979, 1985). This approach arrives at average net rates of generation and destruction of global tectonic environments which can be compared to our data on the age distribution of tin ore deposits (Fig. 4). The metal distribution for tin ore deposits in active margin settings follows approximately

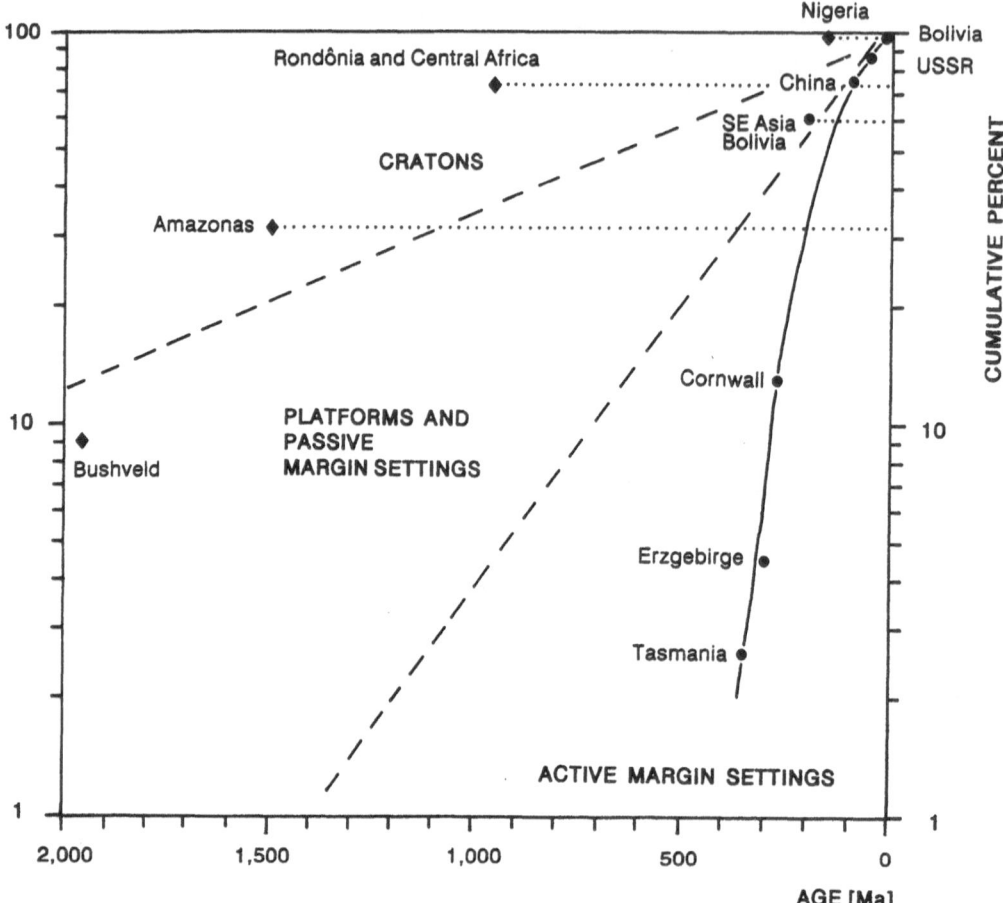

Fig. 4. Age distribution of tin ore deposits (historic production plus reserves; age refers to primary tin accumulation) in intracratonic (solid diamonds) and active margin settings (solid circles). Age domains of major continental geotectonic environments are from Veizer and Jansen (1985) and Veizer (1988)

an exponential pattern, similar to the age distribution of orogenic belts. The half-life (50th percentile) is between 100-200 Ma which implies a survival rate of tin deposits with an age of ≥500 Ma of only 1 % in this setting. The intracratonic environment has a much larger survival rate with a half-life of 690-1800 Ma (Veizer and Jansen 1985), which explains the preservation of Precambrian tin accumulations in this environment.

The geotectonic stability of host environments seems to have generally a first-order control on the time distribution of ore deposits. The statistical investigation by Veizer et al. (1989) arrives at half-lifes for ore deposits in a cratonic setting of ≥1700 Ma for mafic-ultramafic associations, ≥1350 Ma for metamorphic associations, and ca. 800 Ma for volcano-sedimentary associations. This contrasts with half-lifes of hydrothermal ore deposits in orogenic belts (95 ± 41 Ma) and of surficial deposits (a few Ma).

The long-term evolution of the Earth is a combination of both cyclic processes and superimposed unidirectional phenomena. Global chemical trends are a consequence of the crustal growth curve in Fig. 3. Although models of crustal growth differ widely for the early history of the Earth, there is a general consensus that a major portion of the present-day crust was generated until the Archaean/Proterozoic boundary. The early Archaean evolution appears to be dominated by partial melting of mantle material (i.e. basaltic system) with little intracrustal differentiation which would result in the enrichment of the crust relative to the mantle of strongly incompatible elements such as Ag, Pb, Mo, Ta, W, and to a lesser degree Au, Cu, Zn, Sn. Intracrustal melting (granodioritic system) can be expected to become increasingly important with an increase in crustal volume. The granodioritic melt system will result in a separation of the under these conditions compatible elements Au, Ag, Cu, (Zn) from the further incompatible elements Sn, W, Ta, Mo, Pb. Such an interpretation arises from the geochemical data in Table 2, which gives some reference values for major geotectonic units. Primitive mantle (whole mantle plus crust) is computed from analytical data on ultramafic inclusions in the upper mantle (primitive spinel- and garnet-lherzolites), which are taken as being representative for the bulk mantle.

These general observations point to the importance of two metallogenetically critical factors for the formation of tin ore deposits: petrogenesis of tin-bearing granitic rocks (degree of differentiation) and geotectonic position (probability of survival of hydrothermal systems). The following examples focus on the

Table 2. Mean contents of some metals in global geotectonic units of the Earth

	Sn	W	Ta	Mo	Cu	Pb	Zn	Au	Ag
Primitive mantle	0.6	0.02	0.04	0.06	28	0.12	48	0.5	2.9
Bulk crust	1.5	1.0	1.0	1.0	75	8	80	3.0	80
Lower crust	1.5	0.7	0.6	0.8	90	4	83	3.4	90
Upper crust	3.0	2.0	2.2	1.5	25	20	71	1.8	50

Data from Taylor and McLennan (1985), Wänke et al. (1984), Loss et al. (1989). Au and Ag in ppb, all other elements in ppm.

petrological-geochemical evolution of magmatic systems on a regional scale and the associated hydrothermal systems locally developed on this regional framework.

1.4 Global Geochemical Evolution of Tin

Metal enrichment to ore grade is the ultimate outgrowth of efficient large-scale and long-term fractionation processes in a global context (Brimhall 1987). The geochemical evolution of tin and some other elements during the Earth's history is schematically depicted in Fig. 5. The element contents given are derived from analytical data on C1 chondrites, ultramafic nodules from the upper mantle, and on rocks from the continental crust. The bulk Earth composition is from a model calculation by Morgan and Anders (1980) and has a large error margin.

It is generally believed that Type 1 carbonaceous chondrites (C1) closely approximate the condensable fraction of primordial solar system material. The accretion of the Earth appears to have taken place heterogeneously from two chemically different end-member components as a function of oxygen fugacity and temperature (in turn dependent on distance from the centre of the primitive solar system), i.e. a highly reduced component in which all elements with volatilities higher than that of sodium were depleted relative to C1 during earlier cosmochemical fractionation, and a more oxidized component which had a composition close to C1 abundances (Wänke 1981).

Core segregation and ongoing accretion result in an element distribution pattern in the primitive mantle which is controlled by volatility (depletion in volatile elements/enrichment in refractory elements) and siderophility (preferential partitioning of siderophile[*2] elements into the core), i.e. vapour-solid fractionation and metal-silicate fractionation. The volatility of tin

---

[*2] The terms siderophile, chalcophile and lithophile describe those elements that preferentially enter metal, sulphide or silicate phases, respectively (Goldschmidt 1937). This classification is based primarily on the distribution of elements in these phases in meteorites. The behaviour of the elements in the solar nebula is controlled by their volatility (examples in parentheses): gaseous (H, C, N, O, noble gases), very volatile (Bi, Pb), volatile (Sn, Rb, Cs), moderately volatile (K, Mn), moderately refractory (V, Eu), refractory (Ca, Al, U, La) and super-refractory (W, Zr).

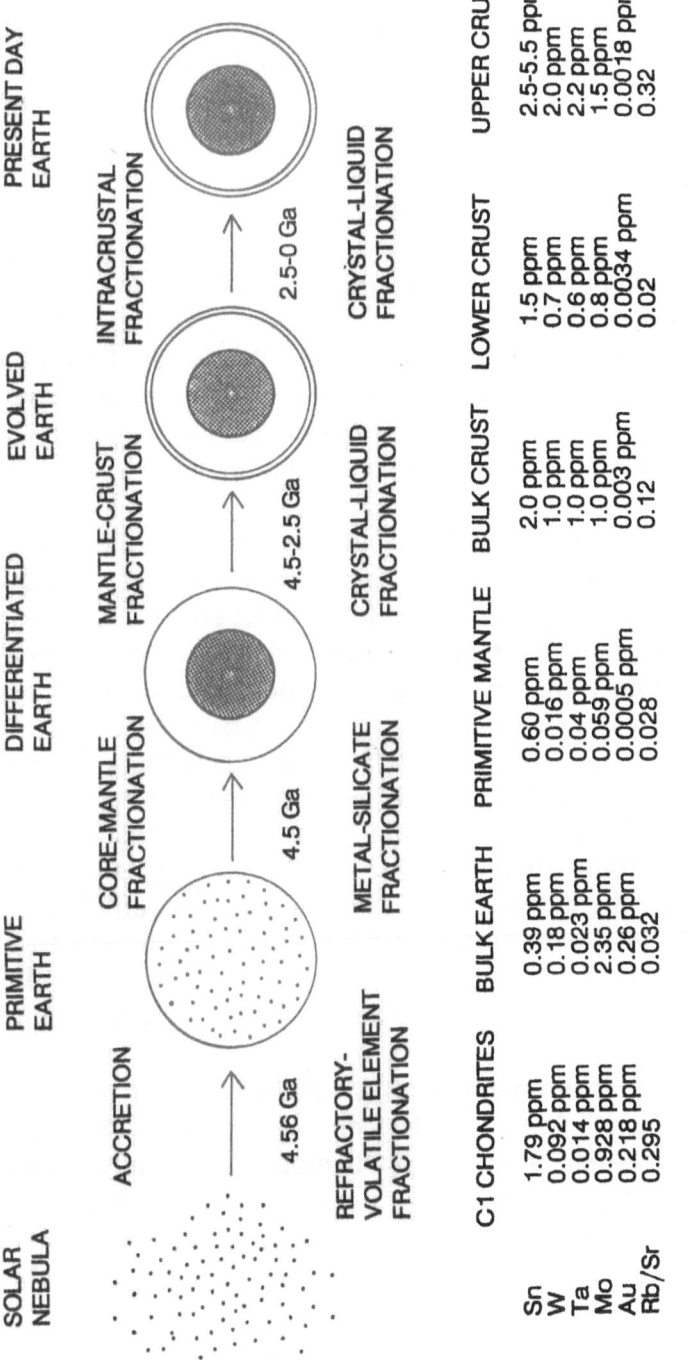

	C1 CHONDRITES	BULK EARTH	PRIMITIVE MANTLE	BULK CRUST	LOWER CRUST	UPPER CRUST
Sn	1.79 ppm	0.39 ppm	0.60 ppm	2.0 ppm	1.5 ppm	2.5-5.5 ppm
W	0.092 ppm	0.18 ppm	0.016 ppm	1.0 ppm	0.7 ppm	2.0 ppm
Ta	0.014 ppm	0.023 ppm	0.04 ppm	1.0 ppm	0.6 ppm	2.2 ppm
Mo	0.928 ppm	2.35 ppm	0.059 ppm	1.0 ppm	0.8 ppm	1.5 ppm
Au	0.218 ppm	0.26 ppm	0.0005 ppm	0.003 ppm	0.0034 ppm	0.0018 ppm
Rb/Sr	0.295	0.032	0.028	0.12	0.02	0.32

Fig. 5. Fractionation pattern of tin and some other elements during the Earth's history. C1 data are from Anders and Grevesse (1989) and Loss et al. (1989). The bulk Earth data are from Morgan and Anders (1980) and are calculated assuming chemical equilibrium between mantle and core. Primitive mantle data are from Anderson (1983). Wänke et al. (1984) and Taylor and McLennan (1985), data on continental crust from Taylor and McLennan (1985)

(50% condensation temperature at 10^{-4} bar total pressure: 720 K) is distinctly different from the volatility of the ore-paragenetically related elements W, Mo, or Ta (50% condensation temperatures at 10^{-4} bar total pressure: >1400 K; Ringwood 1979:100). This results in a drastic decrease of tin content from ca. 1.8 ppm in condensed solar material (as seen in C1 chondrites; Loss et al. 1989) to ca. 0.4 ppm in bulk Earth and 0.6 ppm in primitive mantle (lithophile behaviour). The refractory and moderately siderophile elements W and Mo partition preferentially into the core, which leaves the primitive mantle depleted in these elements when compared to solar composition. However, as deduced from the abundances of siderophile elements in the mantle, general chemical equilibrium between the Earth's mantle and largely metallic core is not established, which is the reason for a large error margin for all bulk Earth chemical models (Newsom and Palme 1984; Wänke and Dreibus 1988). The refractory and lithophile Ta is affected neither by volatile depletion nor by substantial partitioning into the core, and during both cosmochemical and geochemical evolution follows a continuous enrichment trend.

Archaean crust formation by crystal-liquid fractionation of mantle material and dominance of intracrustal fractionation in the Phanerozoic lead to the present-day tin distribution with 1.5 ppm Sn in the lower crust and 2.5-5.5 ppm in the upper crust. The upper crustal range in tin contents is an artifact of analytical uncertainties which arise from conflicting data. The conventional value of 2.5 ppm Sn as compiled by Rösler and Lange (1976) has been revised to 5.5 ppm Sn by Taylor and McLennan (1983) on evidence based on spark-source mass spectrometry. However, this analytical method depends on theoretical sensitivity factors, and new data by a mass spectrometric isotope dilution technique (Loss et al. 1989) confirm the older data and suggest a systematic error by the factor of 1.5-2 in the data of Taylor and McLennan (1983, 1985). The best estimate for the upper crustal tin content is therefore around 3 ppm. The value of 2.32 ± 0.04 ppm Sn for international USGS rock standard BCR-1 (Loss et al. 1989) may be used to calibrate tin data from different laboratories.

It is interesting to note that long-term geochemical fractionation restores in the Earth's crust approximately that tin level which was present in the condensed solar starting material before the onset of cosmochemical fractionation. The same applies to the global fractionation pattern of molybdenum, whereas tantalum provides an example of a continuous cosmo-geochemical enrichment pattern.

1.5 Geochemical Specialization of Tin Granites

The petrological particularity of tin granites was noted very early by Beaumont (1847) on the edge between fact and intuition. The systematic geochemical campaigns in the USSR provided Barsukov (1956, 1957) with a data base to define the geochemical specialization of tin granites as opposed to non-tin granites based on whole rock tin content:

"The investigations carried out have established that granitic massifs not associated with tin ores, while not differing substantially in age or mineralogical composition from tin-bearing granites, do contain tin in amounts of 3-5 ppm, which is equivalent to the Clarke of tin. Massifs, carrying tin ores in varieties unaltered by postmagmatic or contact-metamorphic processes, contain tin in somewhat larger amount - four to five times the Clarke of tin or 16 to 30 ppm (usually 18-26 ppm). Consequently, in the case of tin-bearing granites we can speak of a specialization of the granitic magma from which a given intrusive rock has solidified. Tin-bearing granites are characterized by higher tin contents in those varieties that have not been altered by contact- or postmagmatic processes" (Barsukov 1957:41).

This simple but essential relationship provides the starting point for a qualitatively new period of research on tin deposits, which focusses on geochemical evolution paths in the granitic host environment. Systematic tin distribution patterns allow insight into the evolution of tin during both magmatic and hydrothermal stages and give direct information which is much more obliterated in the comparable environments of copper or molybdenum porphyries.

The geochemical tin specialization of tin granites is accompanied by characteristic enrichment and depletion patterns of other elements. Besides the often anomalously high boron and/or fluorine contents in tin-bearing systems (noted early on by Daubrée 1841), tin granites, when compared to average granitic rocks, are enriched in lithophile elements such as Rb, Cs, Li, Th, U, Nb, Ta and W, and are depleted in elements which are compatible with the granitic main mineral components such as Sr, Eu, Ba, Ti, Co and Ni. Trace element patterns in tin granites as well as the petrochemical equilibration to their high intrusive level at 1 ± 1 kbar (minimum melt composition) point to the important role of fractional crystallization during the magmatic evolution. This situation does not easily allow the application of the usual petrogenetic-tectogenetic classification schemes which are based on source rock chemistry (Chappell and White 1974; Pearce et al. 1984; Brown et al. 1984; etc.).

The geochemical definition of tin granites by Barsukov (1957) based on tin levels is commonly used and mostly valid. A more general definition is preferred here which takes into account the occasionally drastic hydrothermal tin depletion in tin granites: tin granites are granitic rocks for which spatial, temporal and chemical relations point to a causal association with neighbouring tin deposits.

A definition of tin granites must inherently be arbitrary, because those granite phases immediately associated with tin mineralization are parts of much larger composite granitic intrusions of the same petrochemical suite. Granite plutons in tin provinces are circular or elliptical bodies (usually elongated along regional strike) ranging from a few km to several tens of km in diameter, with high intrusion level. They are often assembled in such close proximity that they coalesce or can reasonably be inferred to coalesce into batholiths of regional extent (in tin provinces, for example, the Cornubian batholith, the Erzgebirge batholith, the Main Range batholith in Malaysia, etc.).

In a generalized picture which is in accordance with most situations in tin provinces, the plutons consist dominantly of coarse- to medium-grained K-feldspar megacrystic biotite monzogranite which is intruded by one or several later granite phases with a great textural variety. These subintrusions range from microgranites to granite porphyries and are often characterized by secondary magmatic textures (Cobbing et al. 1986; see Chap. 6.4). Such granite variants occur often in marginal or apical zones within plutons, but may also form the major outcrop area in particular situations with a high erosion level. They are geochemically more evolved than their parental granite host and are affected by processes of pervasive microbrecciation and fluid overprint. Hydrothermal tin-tungsten ore systems are centred on these late granite phases.

2 Petrological Framework

2.1 Magmatic Fractionation

As a consequence of its generally incompatible behaviour shown above, tin becomes enriched in the most fractionated parts of the Earth's crust, i.e. in granitic rocks. The magmatic process of tin enrichment follows, to a first approximation, the law of fractional crystallization (Rayleigh equation). This gives element distribution patterns distinctly different of postmagmatic-hydrothermal tin patterns (Groves 1972; Groves and McCarthy 1978; Boissavy-Vinau and Roger 1980; Lehmann 1982; Higgins et al. 1985; Schermerhorn 1987; Tischendorf 1988). Fractional crystallization in petrology is historically associated with the physical idea of gravitational crystal settling, which, however, need not occur in order for fractional crystallization to proceed. Natural exposures of granitic rocks as well as experimental studies suggest a fluid-dynamically controlled, convective crystal-liquid separation process in the sense of convective fractionation (Rice 1981; Sparks et al. 1984) in which fractionation is due to convection of fluid away from crystals, as opposed to sinking or floating of crystals away from the melt.

Fractional crystallization in evolved granitic rocks results in systematic enrichment and depletion patterns of Ca, Mg, Fe, Ti and many trace elements with only little variation of Si, Al, K and Na. These trends are interpreted to result from sequential crystallization of granitic rock components near the cotectic of the Qz-Ab-Or-An-H_2O system. The conventional reasoning runs along the line that fractionation of mafic minerals, particularly biotite and magnetite, leads to depletion of the residual melt in Fe, Mg, Ti, Co, Ni, Cr, etc.; fractionation of plagioclase results in depletion of Ca, Sr, Eu; fractionation of K-feldspar depletes the residual melt in Ba and Sr; fractionation of the accessory minerals monazite, allanite and titanite depletes LREE in coexisting melt, whereas zircon depletes HREE. Larger, highly charged cations like Sn^{4+}, W^{6+}, Ta^{5+}, U^{4+}, Mo^{6+}, very large cations like Cs^{1+}, Rb^{1+}, as well as small variably charged cations like Be^{2+}, B^{3+}, Li^+, P^{5+}, are incompatible with the major silicate phases (structural constraints) and become enriched in residual liquids.

An alternative explanation for such fractionation trends was advanced in studies on the Bishop Tuff by Hildreth (1979, 1981), emphasizing a process of thermogravitational diffusion. Liquid-liquid fractionation by diffusion of melt components in a thermal gradient (Soret effect) is, however, not compatible with some element relations in fractionation series, particularly with REE patterns (Miller and Mittlefehldt 1984). Evidence for fractional crystallization in the Bishop Tuff has been presented by Michael (1983) and Cameron (1984). Liquid-liquid fractionation is also a process difficult to conceive as an enrichment process for heavy metals such as Sn, Mo and W, which even in simple complexes are relatively heavy with respect to silicate melt compounds. The reduced degree of polymerization in highly fractionated granite melts (i.e. melts rich in H_2O, F and B) would further reduce differences in molecular weights between metal and silicate melt components. Field relations often identify the most fractionated granite phases as being related to small and late-stage subintrusions peripheral to large batholiths, and not limited just to diffuse apical portions of large intrusions, as implied by a thermogravitational model.

This situation can be explained by a model of solidification of large magma chambers from the margins inwards, i.e. sidewall crystallization (McCarthy and Groves 1979; Miller and Mittlefehldt 1984). The relatively cold walls of magma chambers are the most likely place for crystallization to occur. Crystallization on the walls may create a boundary layer of more siliceous and less dense melt which can rise along the edges of the magma chamber, and collect at the top of the system in the form of highly fractionated subintrusions. There must be a transition zone between the largely liquid convecting magma chamber interior and the solid enclosing rock, and the fact of magmatic differentiation points to an efficient crystal-liquid separation process in such a zone. This implies that part of the residual liquid from the crystallizing margins of a magma chamber remains trapped as intercumulus melt whereas another part must return to the interior melt portion. A quantification of such an in-situ crystallization model has been proposed by Langmuir (1989). The consequence of this model is a modification of the $(\bar{D}_i - 1)$ exponent in the Rayleigh equation of fractional crystallization (see below) by a dynamic term dependent on both bulk distribution coefficient $\bar{D}_i$ and the rate of recycled/trapped liquid in the solidification zone, which in turn depends on the physical nature of the individual solidification situation. The complexity of the envisioned crystal-melt separation process during magmatic evolution makes a quantitative modelling on the basis of perfect or equilibrium fractional crystallization questionable (assuming closed or periodically replenished magma chambers). There is the additional complication that there will always be some degree of interaction between the melt system and its wall rocks. The process of combined wall-rock assimilation and fractional crystallization, which is most important for lower crustal environments with a small thermal gradient between magma and its country rocks, has been modelled by DePaolo (1981).

2.2 Geochemical Heritage

The province-bound character of tin deposits and their temporally repetitive nature in some tin provinces ("étagement temporel"; Routhier 1967) provides the basis of the concept of geochemical heritage of tin. A well-known example is the Bolivian tin province, where Precambrian tin mineralization has been thought to be regenerated by Permo-Triassic and Tertiary magmatism (Stoll 1964; Schuiling 1967; Fleischer and Routhier 1970; Schneider and Lehmann 1977). Apparently Proterozoic tin concentrations have been discussed in the Erzgebirge as source of the later Permo-Carboniferous tin mineralization (Baumann 1965; Weinhold 1977). The South China tin province is discussed in terms of a variety of metallogenetic models of heritage which propose synsedimentary tin and tungsten preenrichment from Precambrian up to Devonian times (Liu Yingjun et al. 1984; Cheng Xianyao et al. 1984; Tanelli and Lattanzi 1985; Pei Rongfu and Mao Jingwen 1988).

There is general agreement that granites in association with tin deposits are enriched in a suite of lithophile elements, i.e. elements which are incompatible with the major mineral components during main-stage crystallization. This geochemical specialization can be a result of fractional crystallization, which seems, however, not unequivocally convincing in explaining the tin specialization of tin granites and the existence of tin provinces. Magmatic fractionation is often seen as of minor importance compared to speculative primary tin enrichment in the source material of tin granites or of their wall rocks. Such a model sees the formation of tin granites as a consequence of pregranitic tin enrichments or of regional geochemical tin anomalies (Schuiling 1967; Routhier 1980; Hutchison 1983; and many others).

Geochemical data from granitic fractionation suites allow estimation of both the contribution by pregranitic tin input and the process of intragranitic tin accumulation (Lehmann 1982, 1987). The concept is based on the definition of the partition coefficient $\bar{D}_i$

$$\bar{D}_i = X_i(\text{crystals})/X_i(\text{melt}), \tag{1}$$

where $X_i(\text{crystals})$ and $X_i(\text{melt})$ correspond to the concentration of trace element i in the solid phases (weighted sum of all crystals) in equilibrium with the melt (Neumann et al. 1954). $\bar{D}_i$ is dependent on pressure, temperature and chemical composition of the system.

A melt with total mass M and partial mass m_i of element i develops through crystallization at time $t+dt$ into a residual melt with total mass M-dM and m_i-dm_i, plus a solid mass portion with dM and dm_i. From Eq. (1) follows:

$$dm_i/dM = \bar{D}_i \cdot (m_i - dm_i)/(M - dM) \qquad (2)$$

with $dm_i/dM = X_i(crystals)$ and $(m_i - dm_i)/(M - dM) = X_i(melt)$.

With dmi and dM very small compared to m_i and M, Eq. (2) simplifies to:

$$dm_i/dM = \bar{D}_i \cdot (m_i/M). \qquad (3)$$

Given a crystallization process which prevents continuous reequilibration between solid phases and the melt (slow diffusion in crystal phases, mechanical crystal-melt separation) the recasted Eq. (3)

$$dm_i/m_i = \bar{D}_i \cdot (dM/M) \qquad (4)$$

can be integrated between m(0) and m(t) and between M(0) and M(t) to yield:

$$\log [m_i(t)/m_i(0)] = \bar{D}_i \cdot \log [M(t)/M(0)] \qquad (5)$$

or

$$m_i(t)/m_i(0) = [M(t)/M(0)]^{\bar{D}_i} \qquad (6)$$

Multiplying both sides of Eq. (6) by M(0)/M(t) gives:

$$[m_i(t) \cdot M(0)]/[m_i(0) \cdot M(t)] = [M(t)/M(0)]^{[\bar{D}_i - 1]} \qquad (7)$$

and

$$X_i(t) = X_i(0) \cdot [M(t)/M(0)]^{[\bar{D}_i - 1]}. \qquad (8)$$

The ratio M(t)/M(0) is the fraction of original melt remaining, F, or degree of fractionation:

$$c_i(t) = c_i(0) \cdot F^{[\bar{D}_i - 1]}. \qquad (9)$$

This is the standard equation for all perfect fractional crystallization or distillation processes, known as the Rayleigh fractionation law (Rayleigh 1896; Doerner and Hoskins 1925).

The validity of this relationship for specific sample populations can be examined by log-log variation diagrams of two trace elements i and j, in which linear correlation indicates fractional crystallization. This results from:

$$\log [X_i(t)/c_i(0)] = [\bar{D}_i - 1] \cdot \log F \qquad (10)$$
$$\log [X_j(t)/c_j(0)] = [\bar{D}_j - 1] \cdot \log F. \qquad (11)$$

Equations (10) and (11) combine into:

$$\log [X_i(t)/X_i(0)]/\log [X_j(t)/X_j(0)] = [\bar{D}_i\text{-}1]/[\bar{D}_j\text{-}1] = c. \qquad (12)$$

The constant c corresponds to the slope m of the general linear equation of a straight correlation line with the parameters:

$$\log X_i(t) = m{\cdot}\log X_j(t) + b. \qquad (13)$$

The term b defines the position of the straight line in a log-log plane and can be used as a relative measure of geochemical heritage with respect to any reference system. This is possible by plotting data pairs of a neutral indicator of magmatic fractionation i (such as Ti, Rb/Sr, Zr, Nb, etc.) against element j which is to be tested for primary geochemical specialization. The graphic model of this concept is shown in Fig. 6.

The above theoretical treatment is based on perfect fractional crystallization, i.e. perfect and continuous separation of solid phases and melt. This situation is not realized under natural conditions because there will always be trapped intercumulus liquid in between crystal phases (Langmuir 1989), and because some of the fractionating liquid will move away from the system (Cann 1982). The theoretical model can nevertheless be applied to natural situations because variable amounts of intercumulus liquid modify the degree of enrichment or depletion, but do not change the general log-log evolution pattern (McCarthy and Hasty 1976; Langmuir 1989).

2.3 Crystal-Melt Partitioning of Tin

The main tin carriers in granitic rocks are biotite, hornblende, titanite, ilmenite and magnetite (Barsukov 1957; Petrova and Legeydo 1965; Rub 1968; Tischendorf 1970; Lange et al. 1972). Preferential substitution of the major cations Ti^{4+} and Fe^{3+} by Sn^{4+} is understandable from the similar crystal-chemical parameters of these components [ionic radii (Å) for coordination number VI: Ti^{4+} 0.61, Fe^{3+} 0.55/0.65, Sn^{4+} 0.69; Shannon (1976)]. This explains also the occasionally high tin contents in intramagmatic titanomagnetite ore deposits (105 g SnO_2/mt ore in Grängesberg, Sweden; Schröcke and Weiner 1981:439).

Tin contents of quartz and feldspars reach up to a few ppm, and are always well below the whole-rock tin content. Granitic differentiation suites with little variation in modal composition give linear correlation trends between tin in bulk rock and tin content in individual mineral phases (Rub 1968; Tischendorf

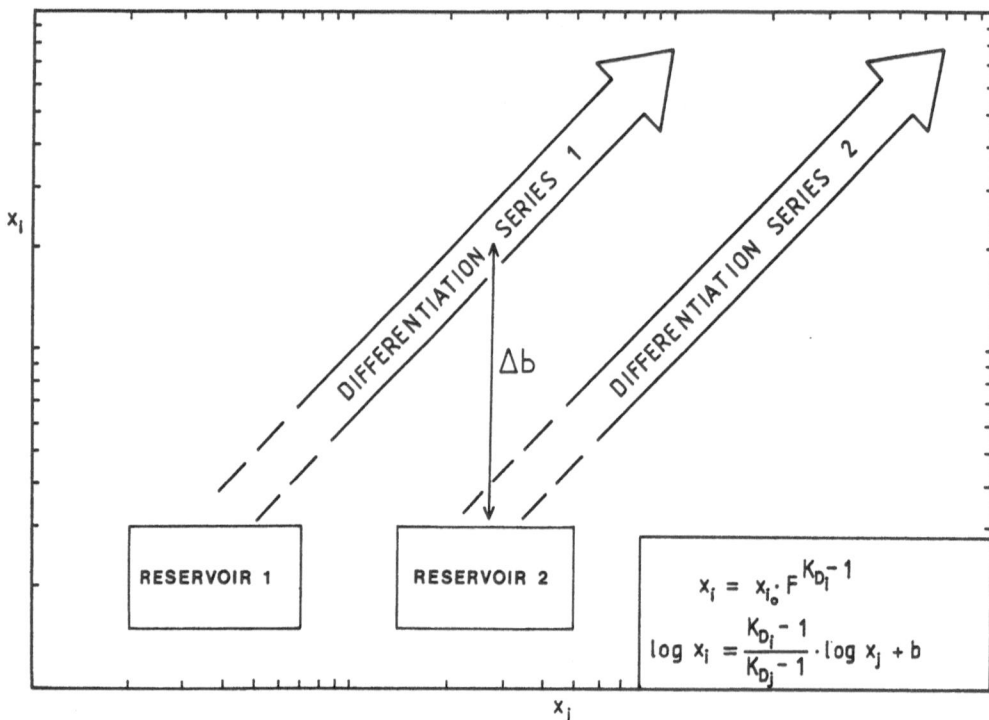

Fig. 6. Quantitative interpretation scheme for geochemical heritage in magmatic fractionation suites. X_i neutral indicator of degree of fractionation; X_j element to be tested for geochemical heritage, i.e. input from geochemically anomalous source

1970; Lange et al. 1972). An example of the relation Sn (biotite)-Sn (whole rock) from the Erzgebirge granites is given in Fig. 7.

A quantitative measure of these distribution trends is given by the tin distribution coefficient D_{Sn}, which in its simplest form and for trace concentrations X_{Sn} can be defined from expression (1) as

$$D_{Sn} = X_{Sn}(\text{crystal})/X_{Sn}(\text{melt}). \tag{14}$$

The bulk tin distribution coefficient $\bar{D}_{Sn}$ for the whole rock is then

$$\bar{D}_{Sn} = [X_{Sn}(i){\cdot}M_i/M + X_{Sn}(j){\cdot}M_j/M + ...X_{Sn}(n){\cdot}M_n/M]/X_{Sn}(\text{melt}) \tag{15}$$

where i, j, ...n are individual mineral phases with mass M_i, M_j, ...M_n of the total mass M of all solid phases.

This definition of the distribution coefficient does not take into account the temporal evolution of crystallization and the resulting heterogeneous element distribution within individual crystal phases. The commonly adopted method

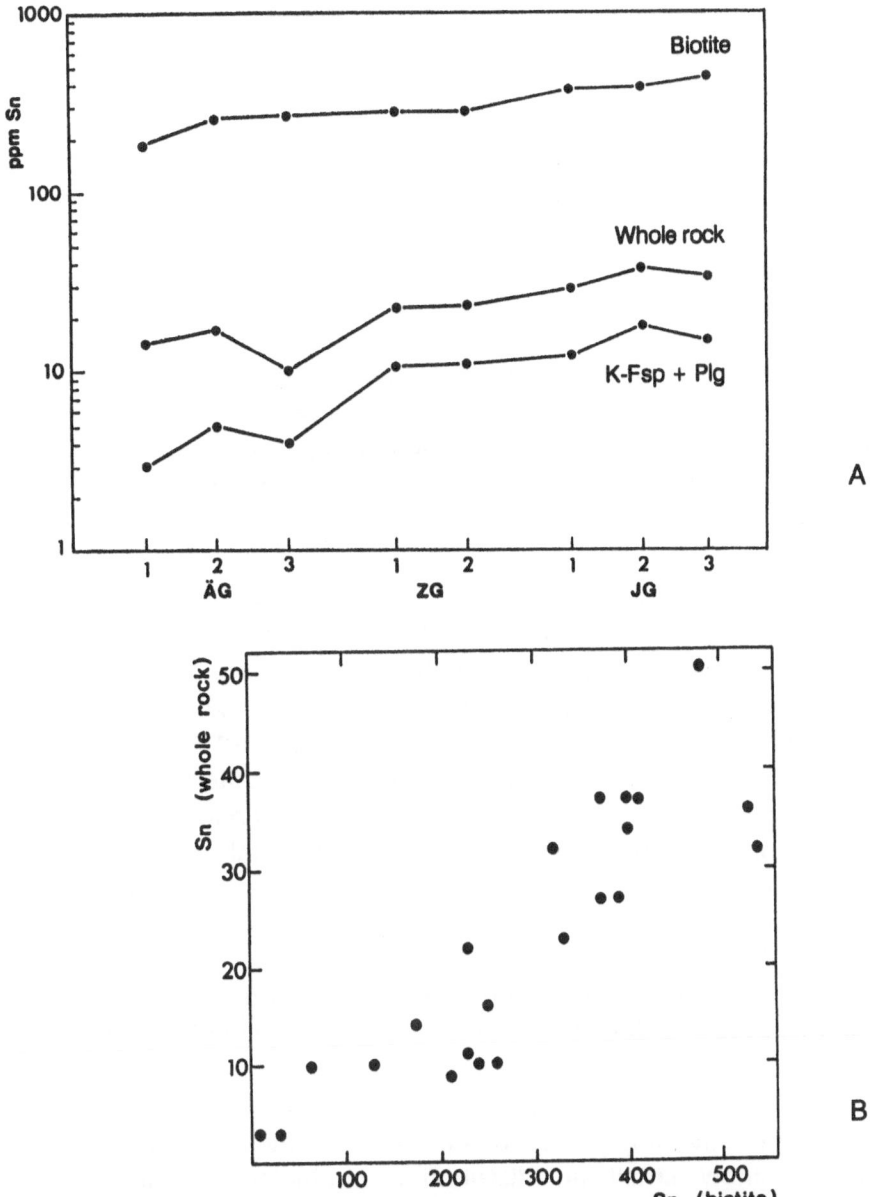

Fig. 7. A The distribution of tin in feldspars (K-feldspar + plagioclase), biotite
and whole rock for individual granite phases of the western
Erzgebirge batholith. ÄG   Ältere Granite (Older Granites); ZG
Zwischengranite (Transitional Granites); JG   Jüngere Granite
(Younger Granites).
B The correlation of tin content in whole rock and in biotite in granite
samples from the Erzgebirge and the Thuringian Forest. Data from
Bräuer (1970), Tischendorf (1970), Lange et al. (1972)

of empirical determination of distribution coefficients by comparing element contents in phenocrysts and their matrix is therefore only a rough approximation.

Empirical tin distribution coefficients determined by this latter method for quartz, K-feldspar, plagioclase, clinopyroxene, olivine, biotite, muscovite and magnetite in various volcanic and subvolcanic rocks of basaltic to rhyolitic composition are compiled in Fig. 8 according to Kovalenko et al. (1988). The data are mean values from a large data base; temperature estimates are derived from petrological equilibria. The corresponding data for tungsten, which give less pronounced trends than for tin, are included in Fig. 8 for comparison. Distribution coefficients for titanite were not measured by Kovalenko et al. (1988). Data from Petrova and Legeydo (1965) suggest for titanite a tin distribution coefficient approximately ten times larger than for magnetite.

The T-D_{Sn} systematics of Fig. 8 allows some conclusions on the behaviour of tin during melt formation. Partial melts with $T \geq 1000\ ^{\circ}C$, i.e. the temperature conditions relevant to melting of mantle material, cannot be rich in tin because such melts are in equilibrium with olivine and/or pyroxene with both extremely low tin contents of less than 1 ppm (Hamaguchi and Kuroda 1969) and $D_{Sn} \approx 1$. The calc-alkaline evolution path of andesitic melts seems to be related to fractionation of magnetite (Gill 1978; Osborn 1979) and will therefore produce tin depletion in the melt. The formation of alkali-rich melts by breakdown of biotite/phlogopite at a low degree of melting is equally a process which at high temperature and correspondingly $D_{Sn} \approx 1$ will not give substantial tin enrichment in the melt fraction relative to the source material. The likely fractionation of magnetite during the calcalkaline magmatic evolution will, in addition, produce a drastic reduction of the tin level of primitive melts.

Granitic melt formation and evolution in an intracrustal environment provides a different situation. The relatively low melt temperature and the correspondingly small tin distribution coefficient for the main mineral components plagioclase, K-feldspar and quartz lead to tin enrichment in the melt phase, an effect partially offset by mafic liquidus phases with $D_{Sn}(xtl/melt) > 1$, such as biotite, muscovite, accessory Fe-Ti phases. Muscovite is, however, unstable in fractionating, high-level magma chambers (≤ 1.5 kbar), even at high B and F levels, and must be considered as of sub-solidus formation (Manning and Pichavant 1984). The small D_{Sn} of quartz will lead to tin enrichment particularly in low-temperature, volatile-rich granite systems near the thermal minimum in which feldspars + quartz co-crystallize.

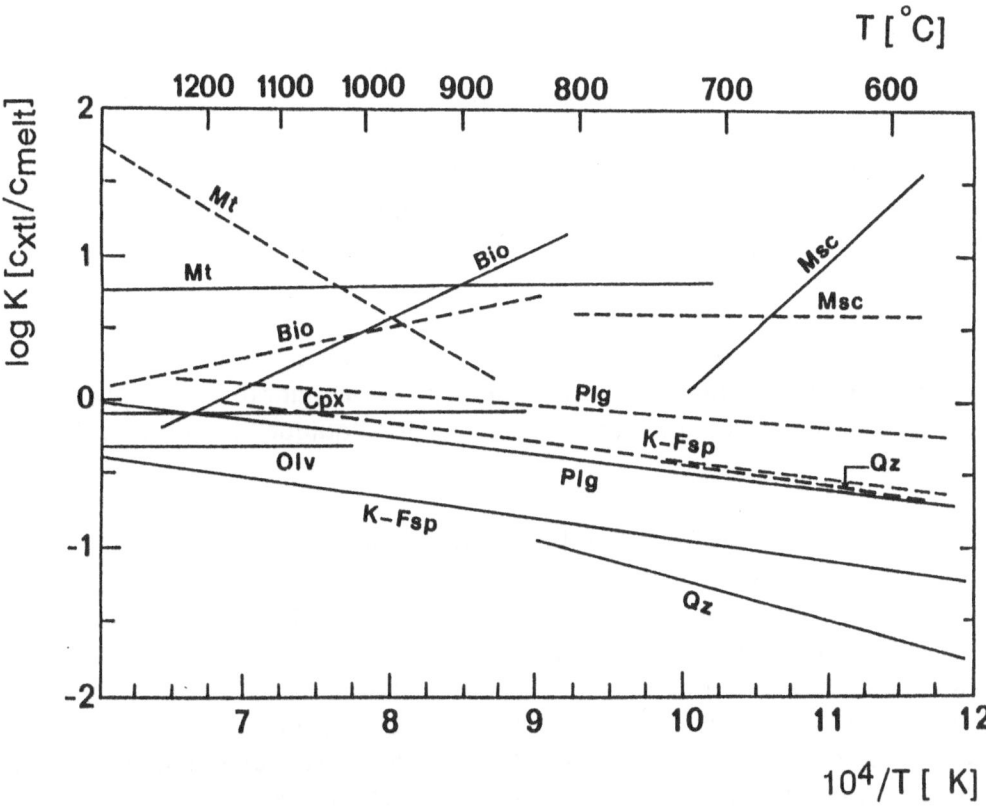

Fig. 8. Tin and tungsten distribution coefficients of magnetite (Mt), biotite (Bt), muscovite (Msc), olivine (Olv), clinopyroxene (Cpx), plagioclase (Plg), K-feldspar (K-fsp), quartz (Qz) as a function of temperature in volcanic/subvolcanic rocks of basaltic to rhyolitic composition. Solid lines are for tin, broken lines for tungsten. Data from Kovalenko et al. (1988)

The bulk tin evolution trend in granitic melts will depend on the proportion of leucocratic/femic minerals. Because the femic minerals crystallize predominantly during relatively early stages of melt evolution, the bulk tin distribution coefficient can be expected to decrease during melt evolution. The increasing degree of fractionation leads to increasing levels of volatile components which have a depolymerizing effect on melt structure. This allows a more efficient fractional crystallization process through lowered melt viscosity (Burnham 1979b) and, in addition, may shift $\bar{D}_{Sn}$(xtls/melt) towards smaller values. Granitic melts can essentially be considered as a mixture of an aluminosilicate polyanionic (network-forming) framework and an interstitial cationic (network-modifying) component. The aluminosilicate structural units form a three-dimensional network composed of polymerized $[AlSi_3O_8^-]_n \cdot [Si_4O_8^0]_m$ units charge-balanced by interstitial cations. The principal depolymerizing components in granitic melts are, besides the ferromagnesian cations (which are drastically reduced during fractional crystallization), alkalies or aluminum in excess of the amount required to charge-balance the alkali-aluminosilicate tetrahedra (i.e. peralkaline or peraluminous melts), as well as anionic depolymerizers of which OH$^-$ and F$^-$ are the most effective (Dingwell 1988). The strongly viscosity-reducing effect of water and fluorine will accelerate crystal-liquid fractionation, extend the temperature range of crystallization, and increase melt component diffusivities, i.e. possible diffusive fractionation mechanisms such as Soret or thermogravitative fractionation (Burnham 1979b; Dingwell 1988). There is evidence that Al^{3+} behaves like a network-modifying cation in peraluminous melts promoting depolymerization of the melt and lowering of the liquidus temperature (Mysen et al. 1985).

Tetravalent tin in silicate melts has probably octahedral coordination with oxygen and is copolymerized with the aluminosilicate network through Sn-O-Si bonds (Stemprok 1989). Increasing concentrations of anionic ligands in the melt which can serve as complexing agents for tin will favour lower crystal/melt partition coefficients. Chloride, fluoride, and hydroxyl strongly complex tin in aqueous solutions. If this relationship holds for silicate melts, an increase of these ligands during fractional crystallization could account for the observed relationship in some rhyolites which show an inverse proportionality between D_{Sn}(xtl/melt) and whole-rock tin content, i.e. degree of fractionation (Antipin et al. 1981; Kovalenko et al. 1984). The mineral-melt distribution coefficients for tungsten have similar characteristics like those of tin (Fig. 8); both temperature and compositional variability are, however, less pronounced than in the case of tin.

Mahood and Hildreth (1983) demonstrated that crystal-liquid partition coefficients in silicic melts can change widely in response to change in melt structure. Urabe (1985) suggested that the alkali/alumina ratio is a major factor controlling activity coefficients of cations in aluminosilicate melts and controlling the metal concentration in released magmatic fluids. Peraluminous melt composition seems to be a precondition for efficient release of chlorine-complexed metals. Partition coefficients are dependent on a variety of melt parameters which are probably not yet all identified, and the presently available specific data allow only qualitative generalizations.

2.4 The Role of Oxidation State

Ishihara (1977) observed two different distribution patterns of accessory opaque minerals in Japanese granitic rocks: a granite population with magnetite, ilmenite, hematite, pyrite and titanite (sphene); and a second granite population with no magnetite and titanite, but with ilmenite, pyrrhotite and ± graphite. Both general granite groups are defined as magnetite- and ilmenite-series, respectively (Ishihara 1977), and differ from each other also by magnetic susceptibility and Fe_2O_3/FeO ratio (both parameters are greater in magnetite-series rocks) (Takahashi et al. 1980; Ishihara 1981). Copper and molybdenum porphyries are part of the magnetite-series granitic rocks, whereas tin granites and porphyries are part of the ilmenite-series granite spectrum (Ishihara 1981).

The different opaque mineral association and Fe_2O_3/FeO ratios in both granite series can be interpreted to reflect different oxygen fugacity during the formation of these rocks, i.e. during their magmatic and subsequent history. Some important oxygen buffers are shown in Fig. 9 as a function of oxygen fugacity and temperature. The separation line of magnetite- and ilmenite-series granitic rocks must be located between the petrogenetic buffer systems hematite-magnetite and quartz-magnetite-fayalite, more specifically near the equilibrium pyrite + magnetite + pyrrhotite and the lower stability limit of the assemblage titanite + magnetite + quartz. The reaction

$$3CaTiSiO_5 + 2Fe_3O_4 + 3SiO_2 \ = \ 3CaFeSi_2O_6 + 3FeTiO_3 + O_2 \quad (16)$$
$$\text{titanite} + \text{magnetite} + \text{quartz} \qquad \text{hedenbergite} + \text{ilmenite}$$

has been calibrated by Wones (1989). The equilibrium expression is

$$\log fO_2 = -30930/T + 14.98 + 0.142(P-1)/T \qquad\qquad (17)$$

where T is the temperature (in K) and P is the pressure (in bar). The equilibrium is affected by impurities in natural occurrences of the above mineral phases. The effect of dilution of the $FeTiO_3$ component by solid solution will be partly offset by solid solution in the magnetite and titanite components (Wones 1989). Natural clinopyroxene-bearing mineral assemblages in granitic rocks will not consist of pure hedenbergite. The effect of a lowered activity of hedenbergite by a greater diopside component is given in Fig. 9. The calculated equilibria correlate with fO_2 reconstructions from the independent Fe-Ti-oxide oxygen geobarometer (see data from Lipman 1971, in Wones 1989).

It is difficult to establish the original oxygen fugacity of a granitic magma from the study of a granite. Slow sub-solidus cooling results in oxidation-exsolution

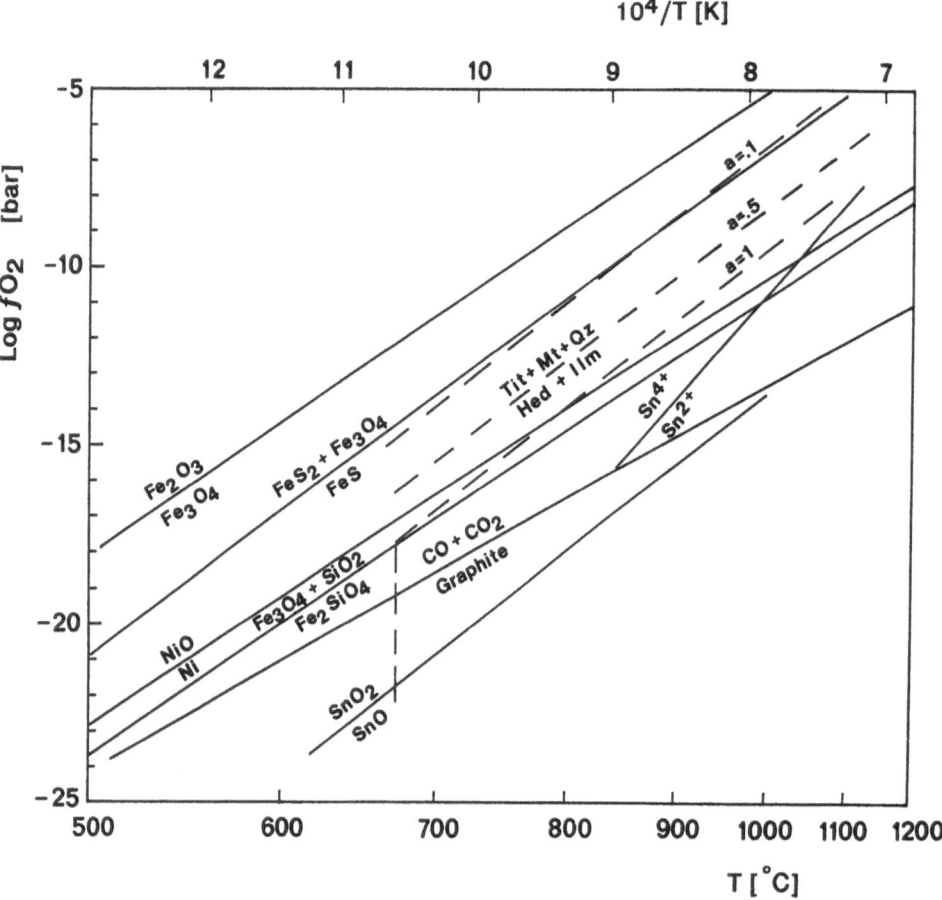

Fig. 9. Some mineral equilibria as function of oxygen fugacity fO_2 and temperature T. Broken lines labelled a = 1, 0.5 and 0.1 depict shift in equilibrium of the assemblage titanite-magnetite-quartz-hedenbergite-ilmenite as a result of lowered activity a of hedenbergite (greater diopside component) in clinopyroxene. Magnetite-series granitic rocks are located between and above the titanite-magnetite-quartz-hedenbergite-ilmenite and pyrite-magnetite-pyrrhotite buffer lines. Ilmenite-series granitic rocks are located below this region with a lower limit at the graphite-CO-CO$_2$ buffer. See text for discussion of Sn^{4+}/Sn^{2+} and SnO_2/SnO equilibrium.
Data sources: fayalite-quartz-magnetite from Hewitt (1978), Ni-NiO from Huebner and Sato (1970), hematite-magnetite from Eugster and Wones (1962), graphite-CO+CO$_2$ at 1 kbar from French and Eugster (1965), titanite-magnetite-quartz-hedenbergite-ilmenite from Wones (1989), SnO$_2$-SnO from thermodynamic data in Hirschwald et al. (1957), Sn^{4+}-Sn^{2+} in sodium silicate from Johnston (1965)

of Fe-Ti oxides which severely limits the application of such phases as an oxygen geobarometer. The recognition of the primary mineral assemblage titanite + magnetite versus clinopyroxene (or amphibole) + ilmenite can therefore be an important aid in reconstructing the relative oxidation state in a melt.

There is also the possibility of inferring the oxidation state of a melt directly from the Fe_2O_3/FeO ratio of the resultant rock. The iron oxidation reaction in silicate melts is commonly expressed in terms of the reaction

$$2FeO + \tfrac{1}{2}O_2 = Fe_2O_3. \tag{18}$$

Experimental data on quenched silicate melts fit an empirical expression of the form (Sack et al. 1980; Kilinc et al. 1983)

$$\ln[X_{Fe2O3}/X_{FeO}] = a\ln fO_2 + b/T + c + \Sigma d_i X_i, \tag{19}$$

where a,b,c and d are regression coefficients and the sum is over the oxide components i. More recent experimental data from Kress and Carmichael (1988) and Mysen and Virgo (1989) confirm that $\ln[X_{Fe2O3}/X_{FeO}]$ is a linear function of $\ln fO_2$ and $1/T$ over a wide range of oxygen fugacities and melt compositions, and that coefficient a in Eq. (19) is 0.21, which leads to the refined iron oxidation reaction (Wones 1988)

$$FeO + 0.232O_2 = FeO_{1.464}. \tag{20}$$

The activity coefficients of the components in Eq. (18) are strongly dependent on melt composition and structure, particularly on the Al/Al+Si ratio and on the NBO/T parameter (non-bridging oxygens per tetrahedrally coordinated cation), which led Mysen and Virgo (1989) to a further refinement of Eq. (19). Given a similar temperature range of crystallization and similar melt composition, $\log[Fe^{3+}/Fe^{2+}]$ ratios in granitic rocks can, however, used as a simplistic approximation of the magmatic $\log fO_2$ conditions. It is assumed that with limited water/rock interaction a rock-buffered external fluid phase will not change significantly the Fe_2O_3/FeO ratio in the rock.

The empirical application of this concept is given in Fig. 10 in which Fe_2O_3/FeO ratios of granitic host rocks from three major ore environments (copper porphyries, molybdenum porphyries, tin porphyries/granites) are compared. SiO_2 is used as the most simple expression for the general degree of magmatic differentiation. The limit between magnetite- and ilmenite series granites is empirically derived (Ishihara et al. 1979). The Fe_2O_3/FeO ratio of this dividing line corresponds closely to the idealized molecular ratio $Fe_2O_3/FeO = 0.5$ for the equilibrium magnetite/ilmenite (assuming no other Fe-bearing phases present).

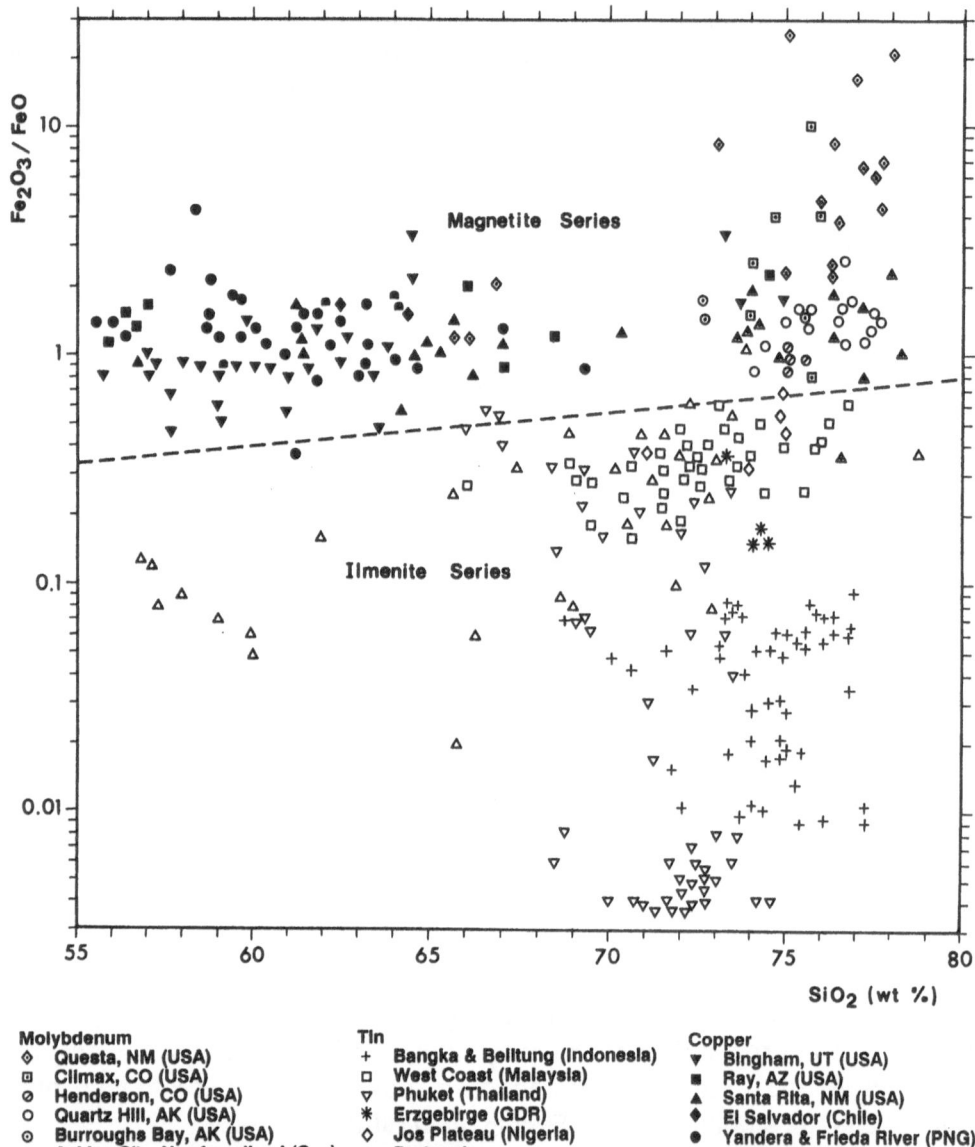

Fig. 10. SiO$_2$-Fe$_2$O$_3$/FeO variation diagram for granitic rocks in association with copper, molybdenum, and tin deposits (copper porphyries, molybdenum porphyries, tin porphyries/granites). The empirical dividing line between ilmenite and magnetite series rocks is for granites in Japan (Ishihara et al. 1979). Rock data compiled from Albuquerque (1971), Banks et al. (1972), Gustafson and Hunt (1975), Hudson et al. (1979, 1981), Imeokparia (1986a), Ishihara (1967), Ishihara et al. (1984), Johnson and Lipman (1988), Jones et al. (1967), Keith (1984), Lange et al. (1972), Lanier et al. (1978), Liew (1983), Mason and McDonald (1978), Moore (1978), Moore et al. (1968), Neiva (1976), Pitfield (1987), Putthapiban (1984), Watmuff (1978), Whalen (1980), Whalen et al. (1982) and White et al. (1981)

The granitic rocks in association with the three major magmatic-hydrothermal ore environments plot in three different fields of the $Fe_2O_3/FeO\text{-}SiO_2$ diagram of Fig. 10. The copper porphyry environment is characterized by relatively little differentiated, magnetite-series granitic rocks. The much more differentiated granitic rocks in association with molybdenum and tin deposits are distinguished from each other by distinctly different oxidation state. Tin granites from the particularly rich tin districts of Phuket (Thailand) and of the Tin Islands (Indonesia) display extremely low Fe_2O_3/FeO ratios.

This general distribution pattern strengthens the assumption that oxygen fugacity plays a critical role in the evolution of these magmatic-hydrothermal systems. Given an effective fractional crystallization process, low fO_2 (i.e. ilmenite-series granites) seems to favour the formation of tin ore systems, whereas high fO_2 (i.e. magnetite-series granites) seems to favour the formation of molybdenum ore systems. The complementary behaviour of tin and molybdenum is also portrayed in the overall bipolar metal distribution of the Sn-W-Mo ore deposit spectrum, with tungsten having a less pronounced affinity for either ilmenite- or magnetite-series rocks.

The redox pair Sn^{4+}/Sn^{2+} provides an explanation for this pattern (analogous: $Mo^{6+}/Mo^{4+}/Mo^{3+}$). The crystal-chemical properties of both tin species are largely different, with ionic radii of $^{VI}Sn^{4+}$ 0.69 Å and $^{VI}Sn^{2+}$ 0.93 Å (Shannon 1976). Under the assumption that both Sn^{2+} and Sn^{4+} species can be stable in granitic melts, a relationship such as $\bar{D}_{Sn^{4+}}$(xtls/melt) >1 (substitution of Sn^{4+} by Ti^{4+} with ionic radius of 0.61 Å at coordination number VI) and $\bar{D}_{Sn^{2+}}$ <1 is required (Ishihara 1981). The theoretical basis of this concept can be formulated for a melt as

$$O^{2-} + Sn^{4+} = Sn^{2+} + \tfrac{1}{2}O_2, \tag{21}$$

and similar to the expression in Eq. (19):

$$\ln[Sn^{2+}/Sn^{4+}] = -\tfrac{1}{2}a\ln fO_2 + b/T + c + \Sigma d_i X_i. \tag{22}$$

Dominance of Sn^{4+} in relatively oxidized melts will result in a bulk tin distribution coefficient $\bar{D}_{Sn}$(xtls/melt) >1, whereas dominance of Sn^{2+} in more reduced melts relates to $\bar{D}_{Sn}$(xtls/melt) <1. The reverse relationship can be formulated for molybdenum.

The thermodynamic background for this argumentation is not well known. The equilibrium $SnO\text{-}SnO_2$ (calculated from data in Hirschwald et al. 1957) is given in the $fO_2\text{-}T$ diagram of Fig. 9. The relation of these components to polymerized tin complexes is, however, vague. The position of the equilibrium far below the QFM buffer indicates that SnO is stable only under very reducing conditions; a situation which is however realized in some basaltic and rhyolitic volcanic rocks (Ewart 1981; Kress and Carmichael 1988; Bryndzia et al. 1989). The experimental determination by Johnston (1965) of the Sn^{4+}/Sn^{2+} equilibrium in $Na_2O \cdot 2SiO_2$ glass at temperatures around 1000 $^{\circ}$C probably provides a better estimate for the speciation of tin in granitic melts. According to these data, divalent tin can be expected to be stable in ilmenite-series granites at a temperature of ≥ 800 $^{\circ}$C (Fig. 9). The data by Johnston (1965) are in agreement with experiments on the valency state of tin in basalt liquids at 1200 $^{\circ}$C which locate the Sn^{4+}/Sn^{2+} equilibrium at log fO_2 -8 bar, with $Sn^{4+}/Sn^{2+} = 0.1$ at log fO_2 -11, and $Sn^{4+}/Sn^{2+} = 10$ at log fO_2 -5 (Durasova et al. 1984). The existence of Sn^{2+} together with the quantitatively dominating Sn^{4+} species in synthetic granite-SnO_2 mixtures prepared under air and at 1600 $^{\circ}$C has been demonstrated by Mössbauer spectroscopy (Sitek et al. 1981).

The experimental investigation of the distribution coefficient of molybdenum between magnetite and silicic melt by Tacker and Candela (1987) confirms the critical role of oxygen fugacity. D_{Mo}(mt/melt) at 800 $^{\circ}$C and 1 kbar with fO_2 one-half a log unit above the NNO buffer was found to be 0.21 ± 0.08. A decrease in fO_2 to the graphite-methane buffer was accompanied by an increase of D_{Mo}(mt/melt) to 0.52 ± 0.13. The same relationship is also valid for the system ilmenite-melt, and a reverse but more moderate trend under the same experimental conditions was found for tungsten (Candela and Bouton 1990).

Besides the role of oxygen fugacity during magmatic tin enrichment, for which experimental evidence is not yet available, the hydrothermal mobility of tin is crucially controlled by fO_2 (see below). The association of tin deposits with low-fO_2, ilmenite-series granitic rocks is therefore of additional importance because oxygen fugacity in a fluid phase will be buffered by the granitic wall rock in an early stage of a hydrothermal circulation system, i.e. at low fluid-rock ratio.

2.5 Solubility of Cassiterite in Silicic Melts

Experimental data on the solubility of SnO_2 in granitic melts have been recently compiled by Stemprok (1990) and are given in a generalized form in Fig. 11. Ryabchikov et al. (1978) first noted the dependence of tin solubility on oxygen fugacity and found a solubility limit for tin in eutectic mixtures of albite-sanidine-quartz-H_2O at 1.5 kbar and 750 $^{\circ}$C of 0.4 wt% Sn at the MW

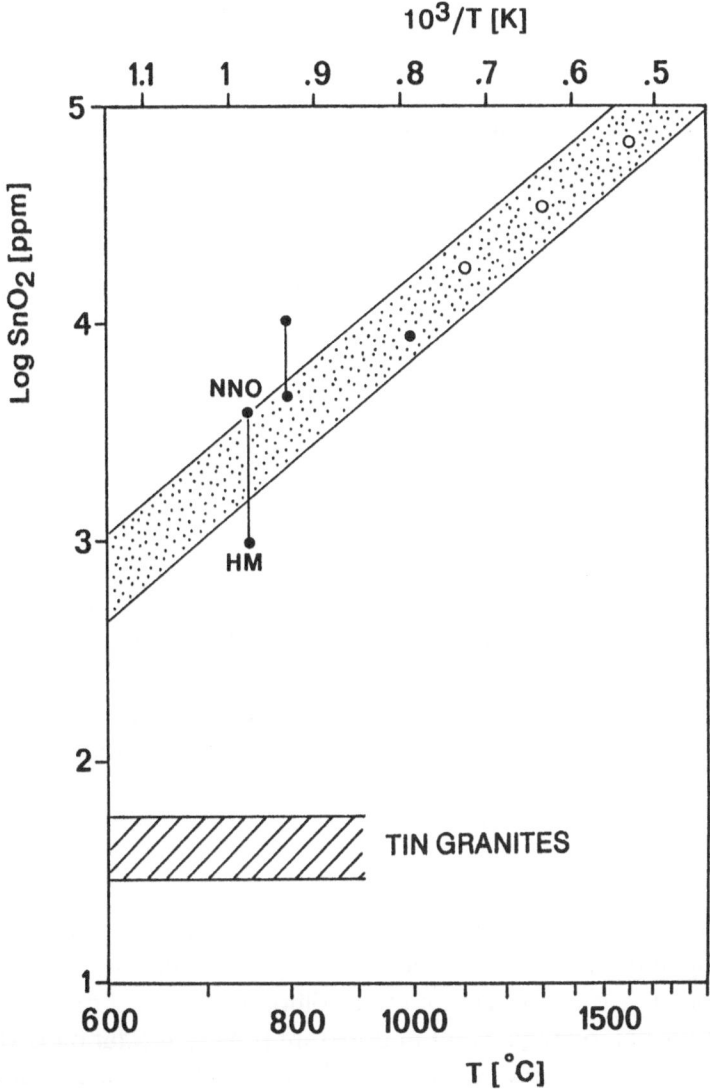

Fig. 11. The experimental solubility of cassiterite in silicic melts as a function of temperature. The stippled trend is from the compilation of Stemprok (1990). Open circles: dry melts at 1 bar and in air; solid dots: melts with excess water at 1.5 kbar and fO_2 between NNO and HM. Data from Ryabchikov et al. (1978) and Stemprok (1990)

buffer, and 0.1-0.2 wt% Sn at the NNO buffer, respectively. These data fit into the general solubility trend for SnO_2 determined at higher temperatures (Fig. 11) and suggest that only in an extremely fractionated environment, i.e. in pegmatites, may a granitic melt become saturated with respect to cassiterite. The average magmatic tin content in tin granites is around 20-30 ppm, much lower than the cassiterite saturation limit. Intramagmatic cassiterite formation is therefore very unlikely even in highly fractionated granites (the often idiomorphic habit of cassiterite may easily lead to minerogenetic misinterpretations).

The components WO_3 and MoO_3 have solubility characteristics very similar to SnO_2 (Stemprok 1990). Both tungsten and molybdenum oxides have a solubility of ≥ 1000 ppm in granitic melts at 750-800 $^{\circ}$C, which only in exceptional situations allows both molybdenite or wolframite/scheelite to be stable liquidus minerals.

2.6 Melt-Fluid Partitioning of Tin

Volatile-rich granitic melts are able to persist to relatively low magmatic temperatures at high crustal levels. The effect of increasing H_2O content on reducing solidus and liquidus temperatures and on phase relations in the experimental system Qz-Ab-Or-H_2O, i.e. expanding the stability field of quartz with displacement of the thermal minimum composition towards the albite corner, has been shown by the classical studies of Tuttle and Bowen (1958) and Luth et al. (1964). The individual effects of fluorine and boron in the same system are similar to water; however, important only with F or B_2O_3 contents in excess of 1 wt% (Manning 1981; Pichavant 1981; Manning and Pichavant 1984).

The experimentally studied volatiles H_2O, F, B and Li have a - possibly additive - fluxing effect on granitic melts, reducing the temperatures of crystallization. The data show that the addition of F and Li to silicic melts has no effect on the solubility of water at constant pressure (Manning 1981; Martin 1983; Webster 1990), whereas the solubility of water increases with increasing B content (Pichavant 1981), an effect also observed with increasing aluminum content in peraluminous melts (Dingwell et al. 1984). Boron-rich and/or peraluminous melts will therefore reach water saturation at a later stage or at lower total pressure than fluorine-dominated and/or metaluminous melts. Little soluble volatiles such as CO_2 or CH_4 will, on the other hand, reduce the solubility of water in a melt (Holloway 1976).

Exsolution of a vapour phase is the necessary consequence of the crystallization of a water-rich melt according to the overall reaction

$$H_2O\text{-saturated melt} \rightarrow \text{crystals} + \text{fluid phase}$$

(Niggli 1920; Burnham 1967; Whitney 1975). Metal partitioning between the melt and fluid phase is dependent on temperature, pressure and melt composition/structure, particularly the activity of complexing agents. The number of controlling variables makes the generalization of specific experimental data and their application to natural systems difficult. There is, however, general agreement that chlorine, the most important metal carrier in aqueous solutions, partitions strongly in favour of the fluid phase (Kilinc and Burnham 1972; Shinohara et al. 1989). A similar behaviour, although less pronounced, is indicated for boron ($D_B$aq/melt $\approx$ 3; Pichavant 1981; London et al. 1988). Fluorine, on the other hand, partitions preferentially into a granitic melt phase relative to a hydrous fluid phase (Hards 1978; Dingwell 1988; Webster 1990). Such partitioning behaviour implies that a cooling water-saturated silicic melt in a closed system will remove fluorine from the coexisting vapour phase and will strongly enrich the same phase in chlorine and boron.

The loss of the fluxing component boron in a boron-rich melt at water saturation results in rapid crystallization concomitant with accelerated release of a fluid phase, which will amplify explosive phenomena in subvolcanic environments. In addition, explosive vapour exsolution will be enhanced through the irreversible reaction of the boron-rich melt with Fe-Mg-bearing wall rock, i.e. tourmalinization (London 1986).

The partitioning of tin in granitic melt-vapour systems has been investigated with very different results. Nekrasov et al. (1982) and Nekrasov (1984) report data from partition experiments of albite-quartz melt with 13.2 wt% Sn and a fluid phase of variable composition (800 °C, 1 kbar, NNO buffer). Their partition coefficients D_{Sn}(fluid/melt) for fluids of various H_2O-HCl-HF-H_3BO_3 compositions scatter widely and are between 0.1 and 0.005 (Nekrasov 1984:97) with a positive correlation of D_{Sn} and chlorine content of the fluid phase. The experiments did probably not attain equilibrium due to low diffusion in the melt (Nekrasov et al. 1982:166) and were possibly hampered by reaction between tin in solution and the container material. In the light of the experiments of Urabe (1985) relatively low partition coefficients for tin may, however, be understandable, taking into account the composition of the synthetic albite-quartz glass charges, i.e. their high alkali/aluminum ratio.

The partitioning experiments by Taylor and Wall (1984) examined the distribution of tin between haplogranite melts and an aqueous chloride-bearing phase with 0-8 m (Na+K)Cl over a range of oxygen fugacities (QFM -1 log unit to HM -1 log unit) at 700-800 °C and 1-3 kbar, utilizing a double gold capsule technique. The exact composition of the "haplogranitic" starting

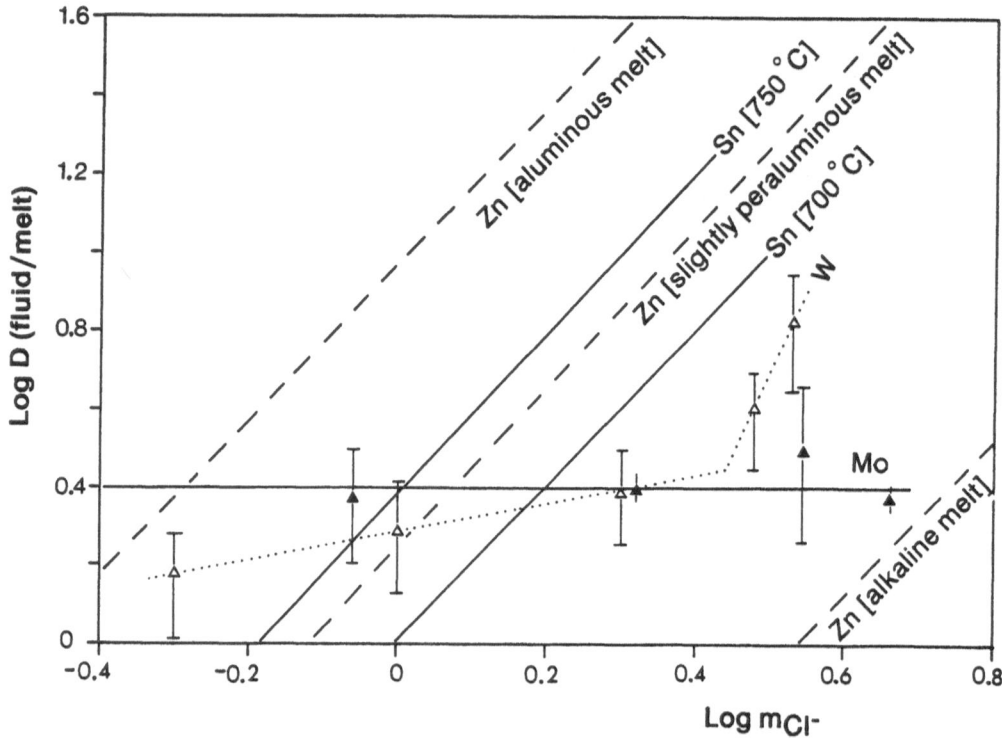

Fig. 12. Vapour-melt partition coefficients for Sn, W, Mo and Zn as a function
of chloride molality. Sn data are from experiments by Taylor and Wall
(1984), complemented by data in Manning and Pichavant (1988), for
700-800 °C, 1-3 kbar, fO_2 fixed but not specified (probably NNO),
haplogranitic melt composition. W data (open triangles) from
Manning and Henderson (1984) for 800 °C, 1 kbar, fO_2 unbuffered,
Qz-Ab-Or-H_2O minimum melt. Molybdenum data (solid triangles)
from Candela and Holland (1984) for 750 °C, 1.4 kbar, melt
composition identical to the experiments by Holland (1972), i.e.
slightly peraluminous granitic melt. Zinc data from Holland (1972) for
830 ± 20 °C, 2.0 ± 0.2 kbar, slightly peraluminous granitic
composition, and from Urabe (1985) for 800 °C, 3.5 ± 0.25 kbar with
both alkaline and aluminous silicic melt

material is not given in Taylor and Wall (1984). Tin was found to partition in
favour of the vapour phase. In detail, the partition coefficient varied as a
function of oxygen fugacity and of the square of chloride molality, with values
of D_{Sn}(fluid/melt) of the order of 1-10. Partition coefficients obtained by
Taylor and Wall (given in Manning and Pichavant 1988) are plotted in Fig. 12
in comparison with data for Zn, Mo and W. The zinc partition coefficients are
included to underline the fact of the extreme importance of melt composition
for such data, with probably the [Na+K+2Ca]/Al ratio of the melt being the
most critical parameter (Urabe 1985). D_{Zn}(fluid/melt) increases for two

orders of magnitude from alkaline to aluminous melt conditions. A similar dependence can be expected for tin which, as deduced from the slope in Fig. 12, is predominantly dissolved as a $SnCl_2^0$ complex at magmatic temperatures, analogous to zinc.

The tungsten data in Fig. 12 show a strong preference of tungsten for the fluid phase. The positive correlation of D_W(fluid/melt) with chloride concentration has been interpreted to suggest varying styles of Cl-complexing, in addition to the probable dominance of iso- and hetero-polytungstates at low chloride concentrations (Manning and Henderson 1984). Because the experiments were run with NaCl solutions, this can, however, also be explained as a result of ion pairing of Na^+ with tungstate, which seems more likely in the light of the lack of correlation between W and HCl concentrations in more recent hydrothermal experiments by Wood and Vlassopoulos (1989). The partition coefficient of molybdenum has been shown by Candela and Holland (1984) to be independent of chlorine (and fluorine) concentration in the fluid, which points to the existence of Mo as molybdate species in the aqueous phase.

Taylor and Wall (1984) conclude from their experiments that the tin-carrying capacity of chloride-rich fluids at magmatic temperatures is of the order of 10^2-10^4 ppm Sn. These results are consistent with hydrothermal solubility data of Haselton and d'Angelo (1986), who found approximately 3000 ppm Sn at cassiterite saturation in 1 m chloride solutions coexisting with a synthetic quartz monzonite (700 °C, 1 kbar, NNO oxygen buffer). The dependence of tin solubility on temperature, chloride concentration and fO_2 as established for magmatic conditions is confirmed by the hydrothermal solubility studies of cassiterite at submagmatic temperatures (Wilson and Eugster 1984; Eugster and Wilson 1985; Eugster 1986).

2.7 Hydrothermal Solubility of Cassiterite

Tin may be present in aqueous solutions in either the divalent or the tetravalent oxidation state, depending on oxygen fugacity and pH of the system according to the equation:

$$H_2O + Sn^{4+} = Sn^{2+} + \tfrac{1}{2}O_2 + 2H^+. \tag{23}$$

The solubility of cassiterite in pure water is very low and, with 1 kbar pressure, 500 °C temperature and oxygen fugacity defined by NNO reaches only 0.4 ± 0.2 ppm Sn, according to the experimental data from Kovalenko et al. (1986).

Solubility data for lower temperatures and with experimental fO_2 conditions not intentionally constrained are slightly lower (Klintsova and Barsukov 1973; Dadze et al. 1981).

Both tin species form aqueous complexes with various ligands, of which Cl^-, F^- and OH^- are the most important in natural hydrothermal fluids (Jackson and Helgeson 1985a). Tin transport via fluoride complexes, first suggested by Daubrée (1849), seems to be less important because fluoride concentrations in natural hydrothermal solutiuons are restricted by equilibrium constraints involving fluorite, topaz, or other fluoride-bearing minerals with relatively low solubilities (Jackson and Helgeson 1985a).

The solubility of cassiterite in HCl-NaCl solutions has been experimentally studied in the temperature range of 400-600 $^\circ$C and at 1.5 kbar pressure by Wilson and Eugster (1984), Eugster and Wilson (1985) and Eugster (1986). For reducing and moderately oxidized conditions (fO_2 fixed by NNO buffer) the dissolution and speciation of cassiterite was found to be controlled by the equation

$$SnO_2 + 2H^+ + Cl^- = SnCl^+ + H_2O + \tfrac{1}{2}O_2 \tag{24}$$

with

$$\log K_{24} = 29.3 - (27500/T),$$

where T is in K. Only under strongly oxidized conditions (fO_2 fixed by HM buffer) does the tetravalent tin species become dominant in the solution, according to the equation

$$SnO_2 + 4H^+ + 3Cl^- = SnCl_3^+ + 2H_2O \tag{25}$$

with

$$\log K_{25} = 33.2 - (18100/T).$$

Chloride complexes of di- and tetravalent tin species are related by an equilibrium analogous to Eq. (23), i.e. higher Cl^- molality, lower pH and higher fO_2 favour tetravalent tin complexes which are, however, important only in very acid solutions throughout the redox stability field of magnetite (Eugster 1986).

If the pH of a solution is buffered by the mineral assemblage K-feldspar-albite-muscovite-quartz (rock-buffered fluid), as is valid for large parts of granitic host rocks of tin deposits, Eq. (24) describes the concentration of the dominant tin complex in a chloride solution. The dramatic dependence of tin concentration on temperature and oxygen fugacity is shown in Fig. 13.

$$SnO_2 + 2H^+ + Cl^- \rightleftharpoons SnCl^+ + H_2O + 1/2O_2$$

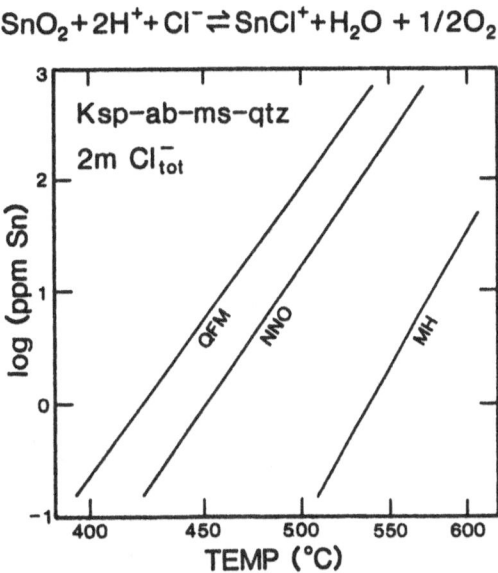

Fig. 13. The solubility of cassiterite in HCl-H$_2$O solutions as a function of temperature and fO_2 defined by QFM, NNO or MH oxygen buffers. The assemblage K-feldspar + albite + muscovite + quartz defines pH, chloride molality is 2 m (Eugster 1986:666)

A solution at a temperature of 500 °C and with stable feldspar will dissolve <1 ppm Sn under conditions of the HM buffer, whereas the same solution under reducing conditions of the QFM buffer is able to transport about 100 ppm Sn. The latter solution will, however, precipitate cassiterite quantitatively in the temperature range 400-450 °C. Tin ore formation below about 400 °C requires a low-pH fluid (or alternatively: high-pH fluid, less probable in natural environments) not in chemical equilibrium with feldspar, in accordance with the common observation of feldspar-destructive hydrothermal alteration with the mineral assemblage muscovite + quartz ± kaolinite around hydrothermal tin-mineralized veins. The solubility of cassiterite under such acid conditions and in the temperature range of 200-350 °C has been calculated by Jackson and Helgeson (1985a,b), taking into account a variety of tin complexes (Fig. 14).

The relatively temperature-independent solubility curve for pH=6 corresponds to the predominance of hydroxide tin complexes with a combined tin concentration around 1 ppm Sn at intermediate pH, in accordance with the experimental data on the SnO$_2$-H$_2$O system (Dadze et al. 1981; Klintsova and Barsukov 1973). At lower pH and even with moderate NaCl concentrations in the fluid, Sn^{2+}-chloride complexes are dominant

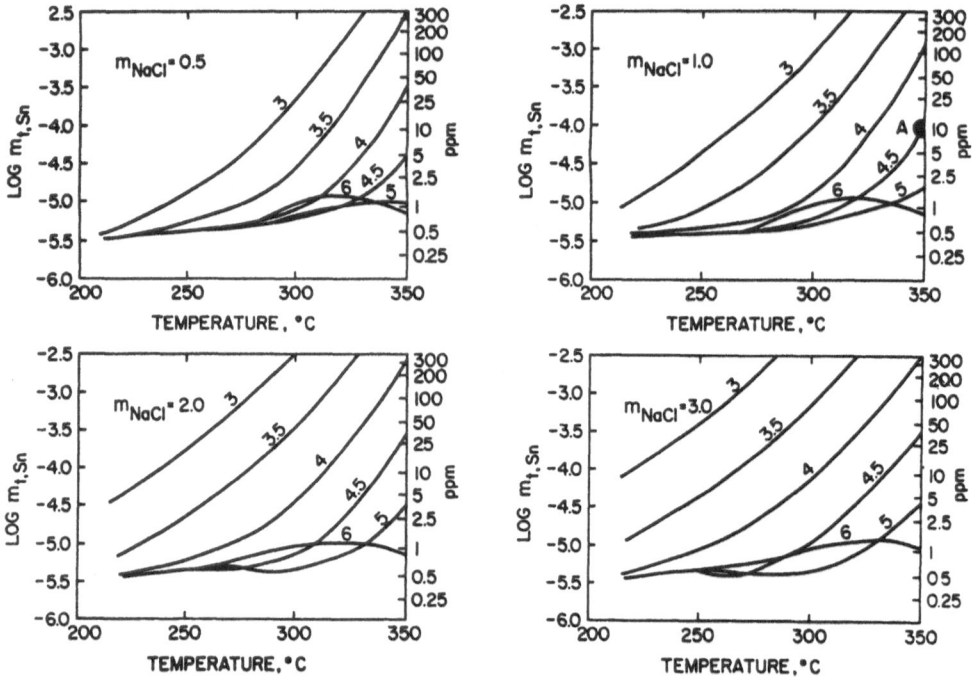

Fig. 14. Solubility of SnO_2 in NaCl solutions in equilibrium with topaz as a function of temperature and pH (numbers on solubility contours). Oxygen fugacity controlled by pyrite-magnetite-pyrrhotite buffer. Pressure along liquid-vapour equilibrium of H_2O (200 °C ≈16 bar; 340 °C ≈146 bar). Total Sn concentrations include: $Sn(OH)^{3+}$, $Sn(OH)_2^{2+}$, $Sn(OH)_3^+$, $Sn(OH)_4^0$, $Sn(OH)^+$, $Sn(OH)_2^0$, $Sn(OH)_3^-$, $SnCl^+$, $SnCl_2^0$, $SnCl_3^-$, SnF^+, SnF^0, SnF_3^-. Point A in diagram for $m_{NaCl}=1$ is an estimate of conditions during early tin ore formation in SE Asian tin belt according to fluid inclusions and mineral association. (Jackson and Helgeson 1985b:1375)

which support tin concentrations in the 100 to 1000 ppm range. Fluoride complexing of tin in such solutions at maximum F⁻ levels allowed by the presence of fluorite or topaz is negligible (Jackson and Helgeson 1985a; Eadington 1988). The steep slope in the diagrams of Fig. 14 points to the importance of temperature control of cassiterite precipitation, which is also dependent on acid neutralization, oxidation and dilution (Cl⁻ concentration). The predominance of hydroxy-tin complexes under alkaline pH conditions is shown in Fig. 15, which indicates a relatively large tin solubility field for pH ≥7 down to moderate temperatures.

Tin ore formation in the epithermal temperature range requires a hydrothermal system partly isolated from mineral buffer assemblages in quartzofeldspathic wall rocks, i.e. a fluid-buffered system (Heinrich and Eadington 1986). This

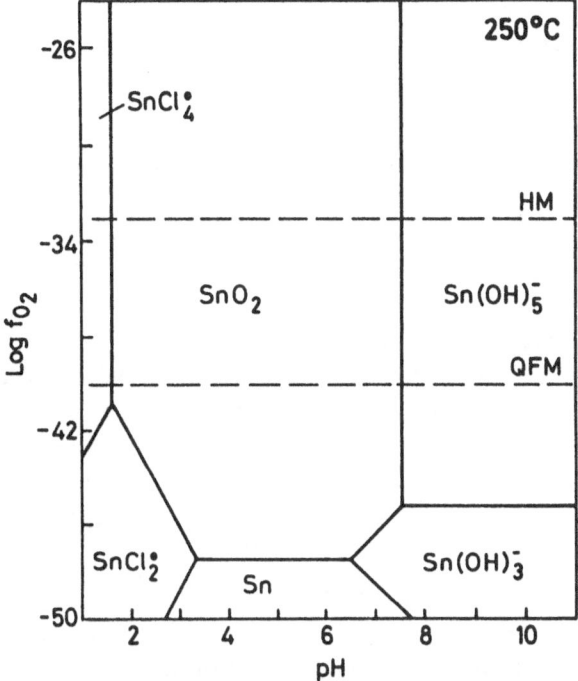

Fig. 15. Equilibrium phase diagram for cassiterite and some chloro- and hydroxy-complexes of tin at 250 °C. Boundaries for the complexes are drawn at an activity of 10^{-3} m, a_{Cl^-} = 1.0. If controlled by the solubility of fluorite, a_{F^-} is too low (2×10^{-3}) for fluoro-complexes of tin to contribute significantly to the solubility of cassiterite. (Eadington 1988:27)

situation is realized once a halo of completely altered wall rock has formed around fluid channels. Subsequent fluid batches will develop extremely low pH and fO_2 conditions with falling temperature and cassiterite precipitation [dissociation of HCl plus chemical effects of Eq. (24)]. Modelling by Heinrich and Jaireth (1989) and Heinrich (1990) indicates that a fluid-buffered solution, instead of precipitating cassiterite, may become strongly Sn-undersaturated as it cools. This allows transport of very high tin concentrations (hundreds of ppm) to a low-temperature deposition site, in which finally acid neutralization through wall rock interaction allows cassiterite mineralization.

A little-studied alternative to hydrothermal tin transport by chloride complexing is a colloidal solution process at low salinity. Occurrences of wood tin ($SnO_2 \cdot nH_2O + SiO_2$) suggest a low-temperature formation from metastable, oversaturated solutions which are able to transport polymerized tin compounds in colloidal form and to precipitate cryptocristalline hydrous cassiterite. Wood tin is known from veinlets in volcanic rocks, relics of colloidal textures are, however, documented also for high-temperature hydrothermal tin deposits (Herzenberg 1936; Lebedev 1967; Seltmann et al. 1985), of which one example is the Chacaltaya district in northern Bolivia (Avila 1982), discussed below.

The hydrothermal solubility of tungsten was recently reviewed and experimentally studied by Wood and Vlassopoulos (1989) under fixed conditions of 500 $^\circ$C and 1 kbar. The average solubility of WO_3 in pure water was around 515 ppm W. Contrasting to the behaviour of tin, no significant increase in solubility was observed in the presence of up to 5 m HCl, implying that chloride complexing of W is not important. WO_3 solubility increased, however, significantly on the addition of NaCl or NaOH, which suggests the importance of a cation-tungstate ion pair such as $NaHWO_4^0$ or polymeric tungsten species. A relatively large scatter for solubility data of WO_3 in low-salinity solutions may be due to the formation of unstable colloids.

The role of chloride seems to be similarly unimportant in the hydrothermal transport of molybdenum. Experimental studies on molybdenite solubility from 200 to 350 $^\circ$C with oxygen and sulfur fugacities fixed by the pyrite-pyrrhotite-magnetite buffer indicate molybdate species to essentielly control Mo transport (Wood et al. 1987). This situation is similar to the partitioning of molybdenum between aqueous vapour and silicate melt, which is unaffected by either chloride or fluoride complexing (Candela and Holland 1984). The hydrothermal mobility of molybdenum in NaCl-KCl-HCl solutions at 300-450 $^\circ$C correlates with oxygen fugacity and pH, i.e. increasing molybdenite solubility with increasing fO_2 and increasing pH (Kudrin 1989). This relationship is inverse to the behaviour of tin.

Note on the concept of oxygen fugacity:

Throughout this text fO_2 is used as the principal redox parameter. It should be borne in mind that dissolved oxygen is a fictive component in most hydrothermal fluids, which conveniently allows to describe the oxidation state of a geological system. The physical presence of dissolved oxygen is not implied in this concept.
A hydrothermal fluid at a temperature of 250 $^\circ$C and with log fO_2 -38, i.e. near the pyrite-pyrrhotite-magnetite buffer, has a molality of oxygen of log m_{O_2} -40.6 (Henry's Law constant of 21000), which is 1 molecule of oxygen for each $10^{16.8}$ kg of fluid. That is less than 1 molecule per ten thousand cubic kilometres of fluid (see the similar calculation for the BR22 well at Broadlands in Barnes, 1984). Therefore, dissolved oxygen in a reduced hydrothermal environment does not physically exist and has, of course, no role as a redox buffer. Even the much more abundant dissolved hydrogen is a negligible buffer component compared to silicate reactions (Giggenbach 1980).

The chances are small that an igneous rock presently exposed at the Earth's surface might be a true copy of the chemical inventory of its past magmatic state. Chemical interaction with a fluid phase during crystallization, cooling and at later stages is inevitable, and is the condition for any hydrothermal ore formation in association with igneous rocks, even though sample suites from igneous rocks often display a more or less extensively preserved magmatic distribution pattern which can be identified because of systematic trends between mineral phases or between individual rock portions. Linearly correlated log-log trace element distribution patterns in granitic fractionation suites, as discussed above, can be interpreted as a result of magmatic evolution and will be increasingly disintegrated into scatter patterns with increasing fluid overprint.

The dominantly magmatic (in a relative sense) distribution pattern of tin in granitic rock suites is shown in the following cases as a function of degree of fractionation. Titanium content and Rb/Sr ratio will be frequently used as an index of fractionation in these examples. Ti has compatible behaviour in granitic melts (although incompatible in mafic melts) and usually has little mobility under hydrothermal conditions. Rb and Sr are easily mobile, but can, however, remain fixed in incipient stages of hydrothermal alteration (low water/rock ratio) by microscopic to submicroscopic blastesis of sericite/muscovite or epidote/carbonate. Such a situation is indicated by a relatively undisturbed Rb-Sr pattern (linear log-log correlation) and systematic Rb/Sr variation complementary to TiO_2.

Zirconium is occasionally used as an additional indicator of fractionation. This element is present nearly exclusively in zircon. Correspondingly, Zr content in a melt is controlled by zircon solubility which is dependent on temperature and on melt composition as expressed by the parameter $(Na+K+2Ca)/Al \cdot Si$ (Watson and Harrison 1983). Zircon saturation in moderately peraluminous granitic melt is around 1250 ppm Zr at 1000 °C, and around 50 ppm Zr at 700 °C. Accordingly, there is a tendency of Zr to become enriched in the melt at high temperature (partial melting and early magmatic evolution), which turns around towards lower temperatures. The change from incompatible to compatible behaviour, and the kinetic problem of metastable pre-magmatic zircon in some granites complicate the interpretation of Zr data in fractionation suites.

The following examples give tin distribution trends for granite suites from a few tin provinces only (Erzgebirge, Malaysia/Thailand, Nigeria). Further data in the framework of the same interpretative model are for the Central African tin province in Lehmann and Lavreau (1987, 1988), for Portugal in Neiva (1984), and for the Bushveld granites in Lehmann (1982), based on data in Lenthall and Hunter (1977). Systematic tin data on the Cornwall tin granites are not available. Some examples of granitic fractionation suites from areas with little or no tin mineralization are briefly discussed below (Nova Scotia, Cape/RSA, SE Australia). Further tin data on non-tin granites are in Biste (1979) for Sardegna, Italy, in Speer et al. (1989) for South Carolina, USA, and in Grohmann (1965) for Austria, among others.

3.1 Erzgebirge/Krusné Hory, Germany and Czechoslovakia

The Erzgebirge or Krusné Hory (German and Czech = ore mountains) is the birthplace of modern mining geology (Agricola 1546) and was long rated as a standard ore province. Silver mining in the Freiberg polymetallic veins started in 1168, and production of tin from placers near Krupka (Graupen) is even somewhat older. However, the historic Erzgebirge tin production is relatively small and corresponds to only about 15 % of the cumulated Cornwall tin output. The great scientific tradition of the Erzgebirge, together with detailed research and exploration work during the last 25 years, however, make it probably the best-studied tin province in the world. The present state of knowledge on the metallogenesis of the Erzgebirge is condensed in the compilation of Tischendorf (1989).

The Erzgebirge mountain region (Fig. 16) is part of the Saxothuringian zone of the European Variscan orogenic belt. It is a WSW-ENE-running fault block (120 x 45 km large) with a large negative gravimetric anomaly, consisting of a sequence of Proterozoic to Lower Paleozoic metamorphic rocks intruded by Variscan granitic rocks which at depth coalesce into the Erzgebirge batholith. The granitic magmatism is divided into an early Variscan cycle (orthogneisses) and a quantitatively dominating late-Variscan cycle (unfoliated granites). The post-kinematic late-Variscan granites are subdivided into two major granite suites, the Older Granites (OG) and the Younger Granites (YG) (Lange et al. 1972). Tin and tin-tungsten ore deposits are spatially associated with the YG suite only, and are located mainly in apical portions of small stocks and their immediate exocontact (Tischendorf

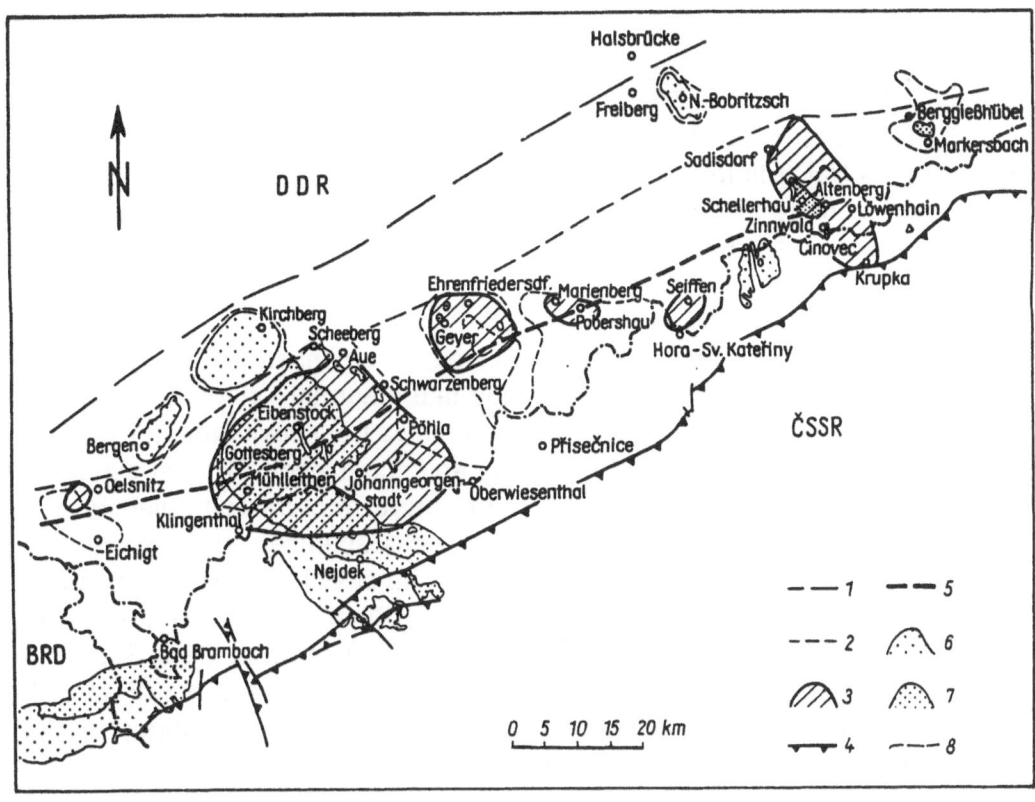

Fig. 16. Generalized geological map of the Erzgebirge tin province (Baumann and Tischendorf 1976:298). **1** NW limit of Older Granite suite; **2** NW limit of Younger Granite suite; **3** areas with tin mineralization; **4** Erzgebirge fault zone (Ohre Graben); **5** axis of tin belt; **6** Older Granites; **7** Younger Granites; **8** depth contour of Erzgebirge batholith at 0 m NN

et al. 1978; Stemprok 1987). Minor tungsten-molybdenum ore occurrences are associated with the OG suite.

Tin and tin-tungsten mineralization is of greisen type (Altenberg, Cinovec/ Zinnwald, Krásno/Schlaggenwald), of stockwork or sheeted-vein/vein type (Ehrenfriedersdorf, Geyer, Krupka/Graupen) and in breccia pipes (Seiffen, Sadisdorf, Gottesberg-Mühlleiten, Sachsenhöhe), with all phenotypes present in variable proportion in each individual ore system. Mineralization of skarn or sulphide replacement type has never been mined, but has a large tin potential (Pöhla, Zlaty Kopec, Halsbrücke). The largest active mine is Altenberg (short of being shut down) which produces 2200 mt Sn per year out of 1,200,000 mt of ore (Mosch and Becker 1985). The Ehrenfriedersdorf, Krásno and Cinovec tin mines produce currently about one tenth each of this figure.

Homogenization temperatures of fluid inclusions in cassiterite define a minimum temperature range of formation of 350-500 $^{\circ}$C; pressures of formation are at ≤ 1 kbar. The ore solutions consist in an early stage of a low-salinity (2-3 wt% NaCl), high-CO_2 fluid phase and a coexisting high-salinity (ca. 35 wt% NaCl) fluid with magmatic stable isotope pattern, and become progressively more diluted by meteoric water during cooling over a period of several million years (Durisova et al. 1979; Thomas and Tischendorf 1987; Thomas and Leeder 1986).

Tin and tungsten ore formation is associated with extensive and widespread hydrothermal overprint, which in an early late-magmatic stage commences with pervasive blastesis of quartz, topaz and mica (muscovite and a variety of dark Li-bearing micas), microclinization and albitization. The subsequent stage of greisen formation is increasingly more fracture-controlled and is accompanied by major metal deposition characterized by the mineral assemblage quartz-topaz-muscovite/zinnwaldite-cassiterite. The large bulk-mining centres of Altenberg (GDR) and Cinovec/Zinnwald (CSSR) with an ore tonnage of 50-100 x 10[6] mt each belong to this type (Cinovec: 0.2 % Sn, 0.035 % W, 0.35 % Li; Altenberg: 0.2-0.3 % Sn, 0.01 % W, 0.01 % Mo, 1 % F; Mosch and Becker 1985; Dasek pers. commun. 1988). Some ore fabrics of the greisen environment are shown in Fig. 17.

The schematic illustration of Fig. 18 demonstrates the zonal arrangement of the mineralogical associations in major types of tin deposits and their spatial relation with adjacent granitic intrusions (Baumann and Tischendorf 1976). The lower parts of the Erzgebirge tin ore systems (endocontacts) are dominated by the topaz-muscovite/zinnwaldite-cassiterite association in a

Fig. 17 (next page). Some textural patterns of greisen-style tin ore of the Ehrenfriedersdorf and Altenberg tin ore deposits.
A Pervasive greisenization (quartz-topaz-mica mineral assemblage) with metasomatic layering in the Ehrenfriedersdorf YG 2 granite. The dark layers consist of Li-mica. Inclusion of greisenized YG 1 xenoclast near hammer (Ehrenfriedersdorf Mine, Sauberg section).
B K-feldspar megablasts with haloes of Li-bearing mica (dark) in greisenized YG 2 granite. Further blastesis of K-feldspar and quartz leads in apical contact zones to the formation of pegmatitic "stockscheider" zones. Length of photograph is about 1 m (Ehrenfriedersdorf Mine, Sauberg section).
C Stockwork/sheeted-vein mineralization in greisenized YG 2 microgranite ("Schnittmuster-Greisen"). Veinlets consist pre-dominantly of Li-mica. Length of photograph is about 1.5 m (Altenberg Mine).

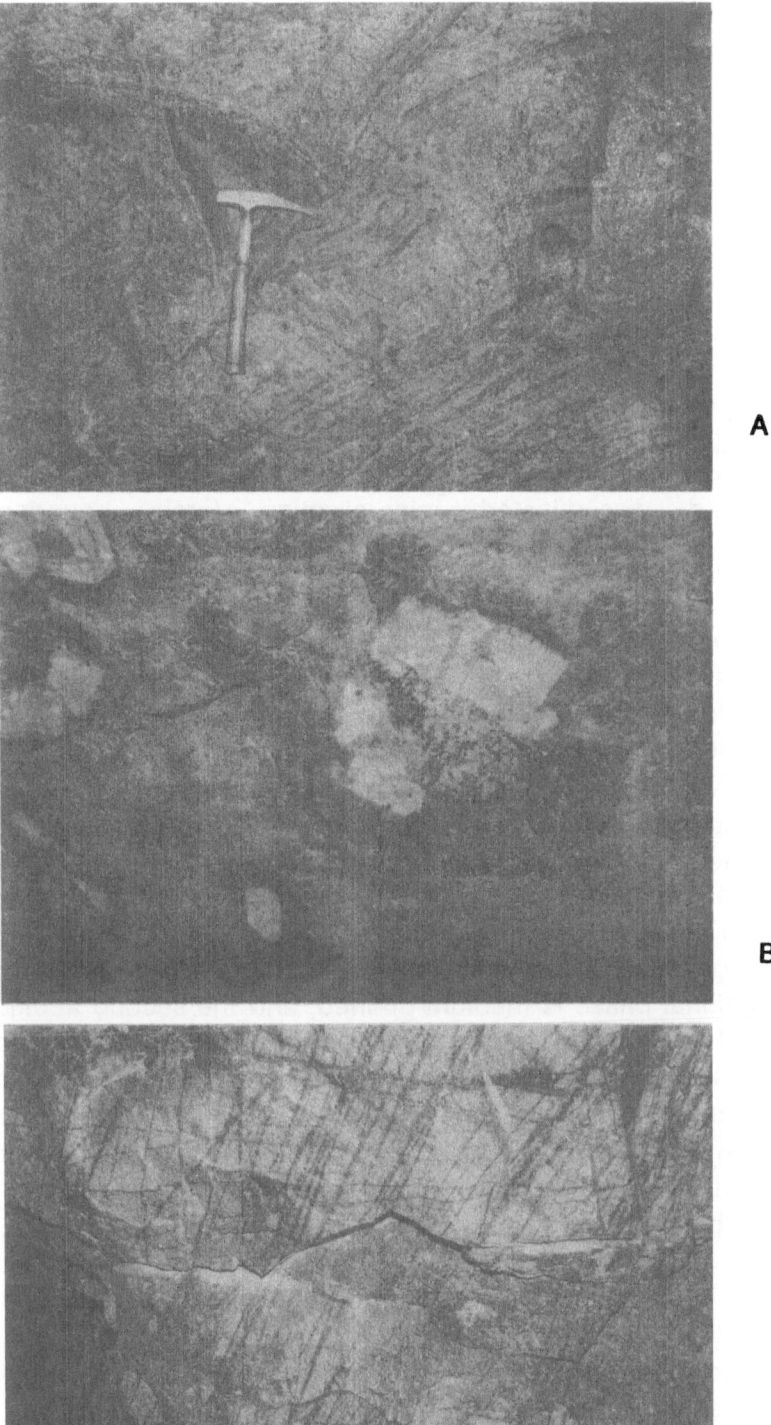

A

B

C

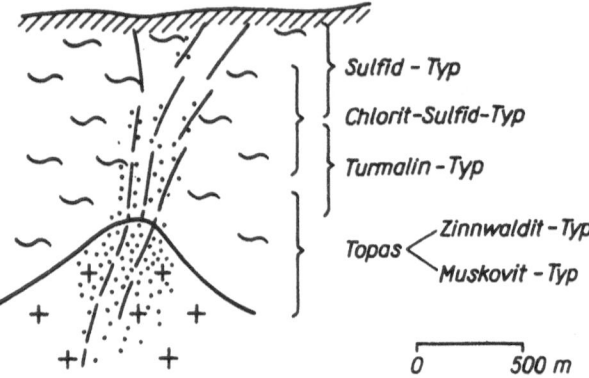

Fig. 18. The zonal arrangement of major morphologic-mineralogical types of tin mineralization in the Erzgebirge (Baumann and Tischendorf 1976:301). Stippled areas denote pervasive hydrothermal alteration and disseminated mineralization (grain boundary-controlled permeability) in conceptual opposition to fracture-focussed fluid circulation and vein mineralization

greisen environment. The exocontact zone has fracture-controlled mineralization which grades from a tourmaline- into a chlorite-dominated mineral association with an increasing amount of sulphide minerals (pyrite, arsenopyrite, chalcopyrite, bismuth, bismuthinite, stannite, etc.).

The OG suite consists of biotite monzogranites, the YG suite is composed of biotite syeno- to monzogranites with substantial amounts of sub-solidus muscovite. Each suite can be divided into a main intrusive phase and two additional successive intrusive phases (OG 1→2→3; YG 1→2→3). The main intrusive phase is distinguished by a coarse-grained porphyritic texture, the first additional phase is medium-grained, and the second additional granite phase is fine-grained. The quantitative proportions of the individual granite phases are approximately 60:30:10 according to outcrop dimensions in the western Erzgebirge (Tischendorf et al. 1987). In an intermediate temporal position in between the OG and YG suites occur locally so-called Intermediate Granites (IG 1→2) and Transitional Granites (OGt) with a distinct texture and composition.

The granitic intrusions appear to form a composite batholith at depth which underlies the entire Erzgebirge block (Watznauer 1954). The country rocks of the granite intrusions consist predominantly of Ordovician phyllites in the western part of the Erzgebirge, and of Proterozoic paragneisses ("Graugneis") in the eastern part. The erosion level in the western Erzgebirge is relatively deep with an exposure of granitic rocks at a plutonic level, whereas the eastern Erzgebirge granites are exposed in most apical portions and at a subvolcanic level (Fig. 19). The eastern Erzgebirge granites and cataclasite-subvolcanic complexes are characterized by multiple episodes of large-scale fluid-explosive brecciation and concomitant greisenization in a caldera setting (Seltmann et al. 1990). Geological and thermobarometric data suggest for the Altenberg tin deposit a depth of formation of 1000-1500 m below the paleosurface, whereas the corresponding figures for the Ehrenfriedersdorf deposit are 2000 m, and for the Eibenstock deposit in the western Erzgebirge 4000 m (Thomas 1982).

The relative age positions of the individual granite phases are well documented by field relationships. There is, however, no plain radiometric evidence for age differences between the OG and YG suites. All granite intrusions appear to have an Upper Carboniferous age in the range of 300-320 Ma (Gerstenberger et al. 1984). Rb-Sr areal isochrons of the OG and YG suites are identical within the analytical error margins and define an age of 317 ± 5 Ma. Sr initial ratios are around 0.707 for the OG suite and only around 0.702 for the tin-bearing YG suite. On the basis of these values, a petrogenetic model has been put forward recently which derives the Erzgebirge batholith from mantle material, with a major degree of crustal contamination in the OG

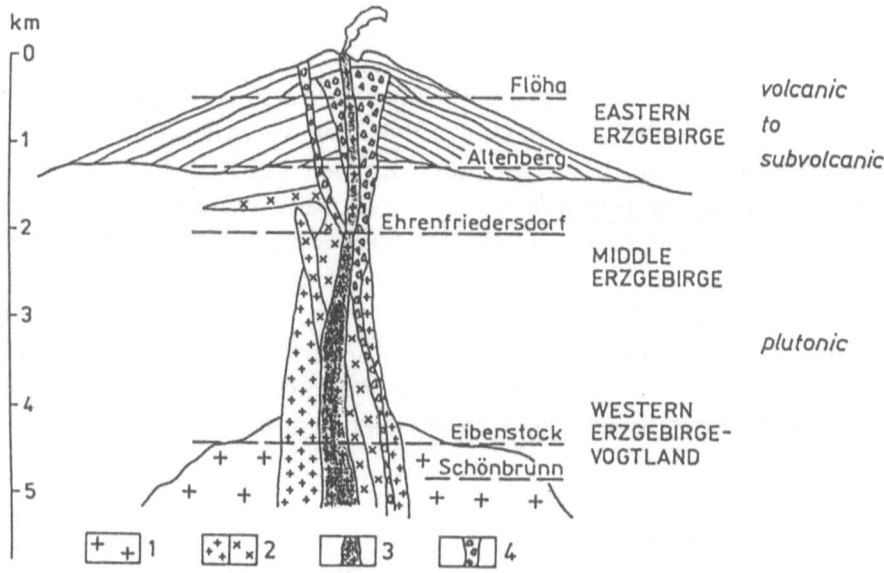

Fig. 19. Structural setting and recent erosion level (broken lines) of Sn-(W) ore deposits of the Erzgebirge according to Seltmann et al. (1989, 1990). 1 Older Granites (OG suite); 2 Younger Granites (YG suite); 3 microgranite dykes; 4 breccia pipes

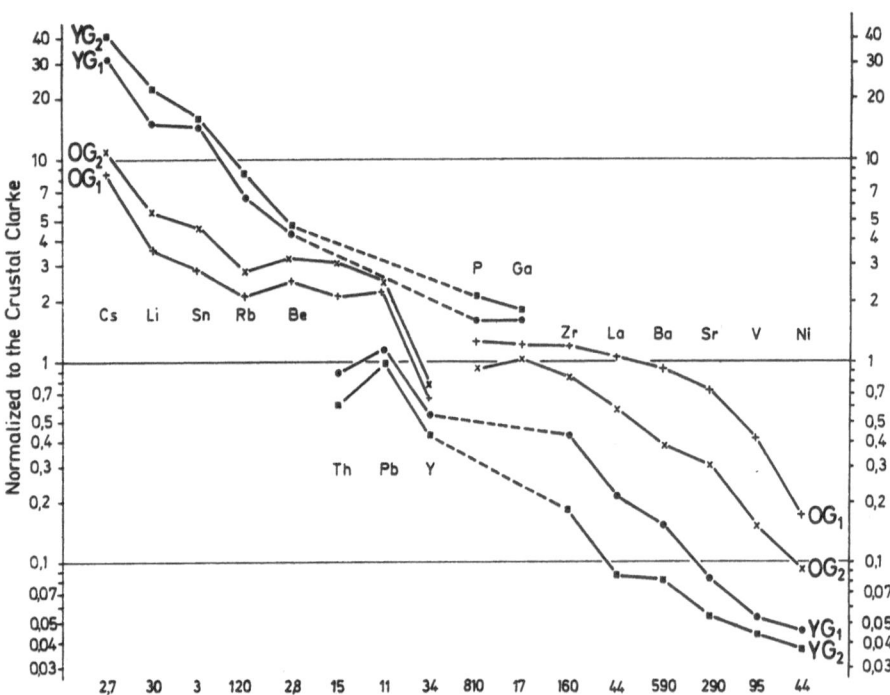

Fig. 20. Multi-element spectrum for major granite phases of the Erzgebirge (OG 1-2 Older Granites; YG 1-2 Younger Granites), normalized to average crustal element contents (CLARKE values) specified in the lowermost column (in ppm). (Tischendorf et al. 1987:227)

suite and less crustal input in the YG suite (Schütze et al. 1984; Gerstenberger et al. 1984; Stiehl 1985; Dahm et al. 1985). The exceptionally low Sr initial ratio of the YG suite may, however, result as well from postmagmatic rubidium metasomatism, which is a very widespread and typical phenomenon in the Erzgebirge tin granites and which has been shown to affect the intrusions up to several Ma after their solidification (Gerstenberger 1989). Such an explanation would allow an essentially identical lower crustal source for both OG and YG suites. First Nd isotope data from the Altenberg and Eibenstock tin granites with initial ϵ_{Nd} values of 0.0 and -6.0, respectively (Gerstenberger 1989), point nevertheless to quantitatively variable involvement of mantle material in these rocks.

Based on numerous Rb-Sr and K-Ar ages and on field relations, the older concept of two major phases of granite magmatism in the time intervals of 340-310 (OG suite) and 305-280 Ma (YG suite) is still widely accepted, with tin mineralization associated with the end of the younger granite cycle (Lorenz and Schirn 1987; Stemprok 1986; Tischendorf et al. 1987). This concept is

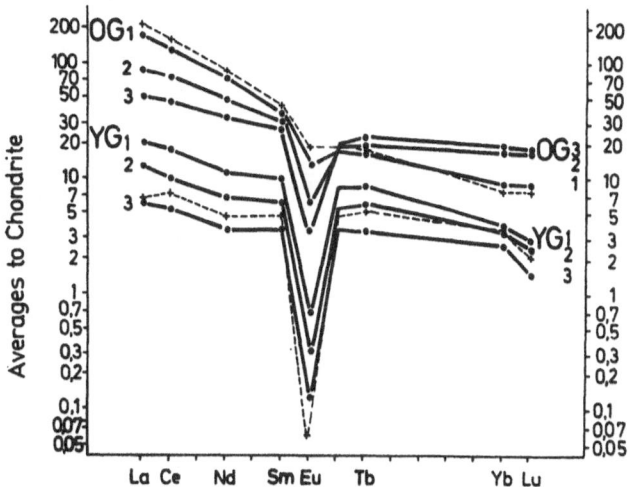

Fig. 21. REE distribution pattern of Erzgebirge granite suites (**OG 1-3** Older
Granites; **YG 1-3** Younger Granites). Arithmetic means of 38 rock
samples; stippled lines are extreme values. (Tischendorf et al.
1987:220)

compatible with the geological evolution in neighbouring Hercynian granite
provinces (Fichtelgebirge, Black Forest and Vosges Mountains, Massif
Central).

Both Older and Younger Granite suites have peraluminous composition, with
mol. Al_2O_3/Na_2O+K_2O+CaO 1.06-1.11 for OG and 1.20-1.27 (muscovitization) for the YG suite. The entire granite sequence from OG1 to YG3
appears to be interlinked by systematic chemical enrichment and depletion
patterns, i.e. successive enrichment in F, Cs, Li, Rb, Ta, Sn, W and
complementary depletion in Ti, Fe, Mg, Ca, Co, Cr, Ni, V, Zr, Sc, Hf, Ba, Sr,
REE (Figs. 20 and 21).

The trace element trends indicate a process of magmatic evolution
predominantly controlled by fractional crystallization. The degree of
fractionation F of the youngest granite phase of the OG suite has been
estimated by Budzinski and Tischendorf (1985) as 0.1-0.2. The YG suite
marks for some elements and isotope ratios a hiatus with the OG suite, and is
also modified by fluid interaction, but its consistent and systematic element
distribution pattern is in favour of an interpretation as the most evolved granite
stage of a general late-Hercynian differentiation suite. The latest YG3
subintrusions consist of small alkalifeldspar aplogranite bodies which display
an extreme degree of fractionation. Ti-Ta data from the Altenberg, Sadisdorf
and Zinnwald alkalifeldspar aplogranite stocks (Just et al. 1987; Tischendorf

54

1989) define a range of 50-200 ppm Ti and 15-100 ppm Ta for both magmatic
and hydrothermally overprinted (mineralized) rocks (Fig. 22). The immobile
and incompatible nature of tantalum allows an estimate of the minimum

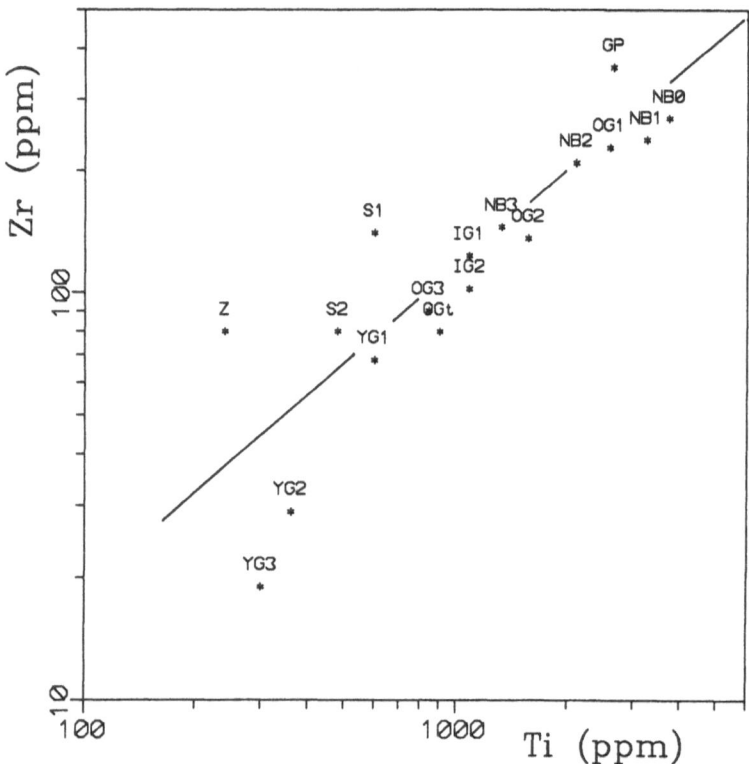

Fig. 22 (continued on next page). Ti-Zr, Sn-Ta and Ta-Ti variation diagrams for
    various granite units of the Erzgebirge. Data points are arithmetic
    means from the compilation of Tischendorf (1989). Correlation
    coefficient r for log[Ti]-log[Zr] is 0.86 (n=17), for log[Ta]-log[Sn]
    0.88 (n=17), for log[Ti]-log[Ta] (n=23) -0.94.
    NB 0-2 Niederbobritzsch massif (least-evolved part of OG suite); **GP**
    Granite porphyry of Altenberg-Frauenstein (OG suite); **OG 1-3** Older
    Granites (Aue, Bergen, Schwarzenberg, Flaje, Kirchberg); **OGt**
    Transitional Granites (Bergen-type); **IG 1-2** Intermediate Granites
    (Krinitzberg, Walfischkopf; xenoliths inside of YG suite); **S 1-2**
    Schellerhau granites (early YG suite); **YG 1-3** Younger Granites
    (Eibenstock-Nejdek, Schellerhau, Altenberg, Zinnwald, Sadisdorf,
    Sachsenhöhe, Greifenstein, Ehrenfriedersdorf, Geyer; **Z** Zinnwald
    alkalifeldspar aplogranite (latest subintrusion of YG suite). Data for
    Ti-Ta diagram include in addition (Just et al. 1987): **Sa 1-2** Sadisdorf
    syeno- and monzogranite; **A 1-3** Granite suite of the Altenberg ore
    deposit (analogous to YG 1-3 suite), with A 3 being an alkalifeldspar
    microgranite; **Ā** locates arithmetic mean of mineralized samples of
    greisenized granite unit A 2

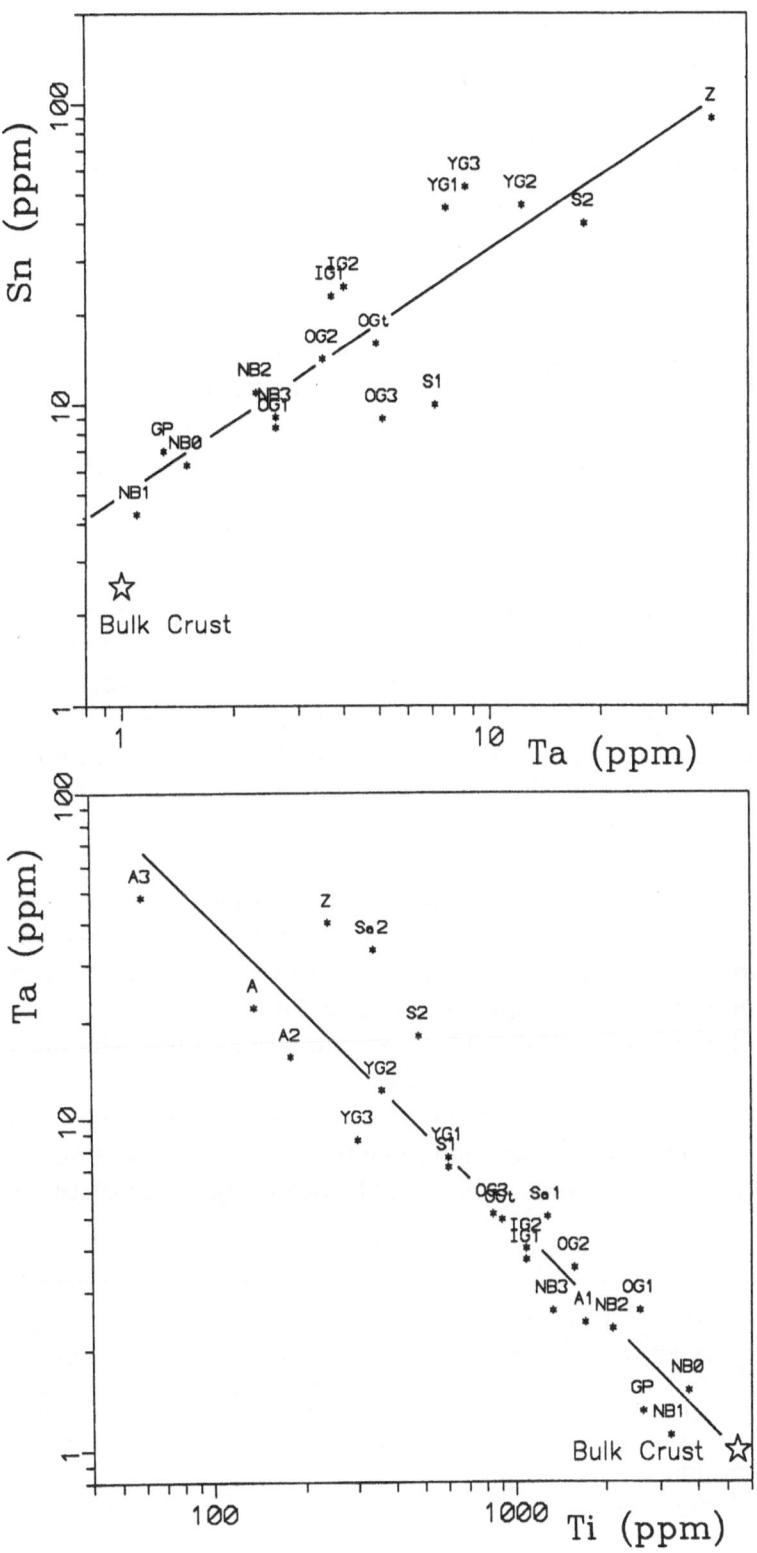

degree of fractionation which gives F = 0.01 for most evolved rock portions (calculated from Eq. 9 in Chapter 2.2 with the limiting assumption of $\bar{D}_{Ta}$ 0).

The distribution pattern of tin as a function of TiO_2, Zr and Rb/Sr is given in Fig. 23. According to our general model depicted in Fig. 6, the contents of titanium, zirconium and the Rb/Sr ratio are taken as three independent indicators of fractionation of granitic melt. The Erzgebirge granite suites follow essentially a tin enrichment pattern in accordance with a fractional crystallization model (linear correlation of trace elements in log-log space). The least evolved granite portions have tin contents of 5-6 ppm, tin levels which would be expected by partial melting of average crustal material. There are no indications of a regional geochemical specialization in tin previous to the large-scale action of magmatic fractionation in the Erzgebirge granite batholith.

The geological situation of the western Erzgebirge is very similar to the neighbouring Fichtelgebirge in eastern Bavaria. The trace element trends of the Fichtelgebirge granite suite look like the Erzgebirge trends, without, however, reaching the very high degree of fractionation typical for the tin-bearing granite phases of the Erzgebirge (Richter and Stettner 1979; Tischendorf et al. 1987) (Figs. 24 and 25). Differentiation suites for individual granite intrusions of the Fichtelgebirge are documented in Richter and Stettner (1979, 1987) and Richter (1984).

The late-orogenic Hercynian granites of the Black Forest and Vosges Mountains are similar to the Fichtelgebirge and Erzgebirge granites as well. Rb-Sr isochron data from the Black Forest and Vosges Mountains document an early-orogenic granite suite of 365-329 Ma in age, followed by post-orogenic biotite granites with 325-310 Ma and biotite-muscovite granites with 300-280 Ma (von Drach et al. 1974; Brewer and Lippolt 1974). Initial $^{87}Sr/^{86}Sr$ ratios of 0.708-0.730 suggest crustal source material. Fractional crystallization controls the magmatic evolution of at least the youngest and most evolved granite phases which reach locally (granites of Bärhalde and

---

Fig. 23 (next page). TiO_2-Sn, Rb/Sr-Sn and Zr-Sn variation diagrams for Erzgebirge granite phases (arithmetic means ± one standard deviation). Trace element data from Tischendorf et al. (1987). Correlation lines are statistically significant at a confidence level of >99 %. Reference fields for global averages of bulk and upper crust according to Taylor and McLennan (1985), shale data from Rösler and Lange (1976)

57

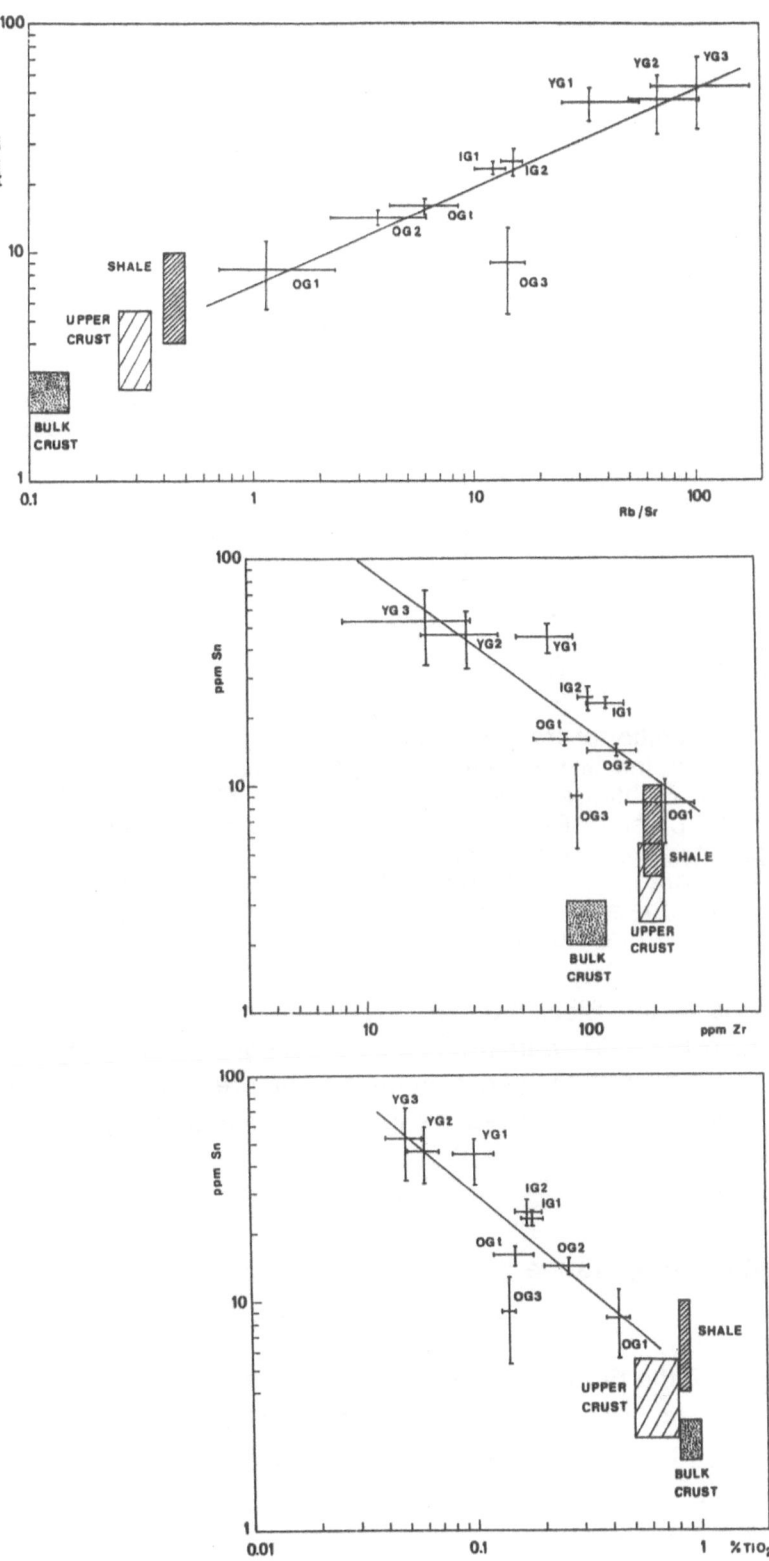

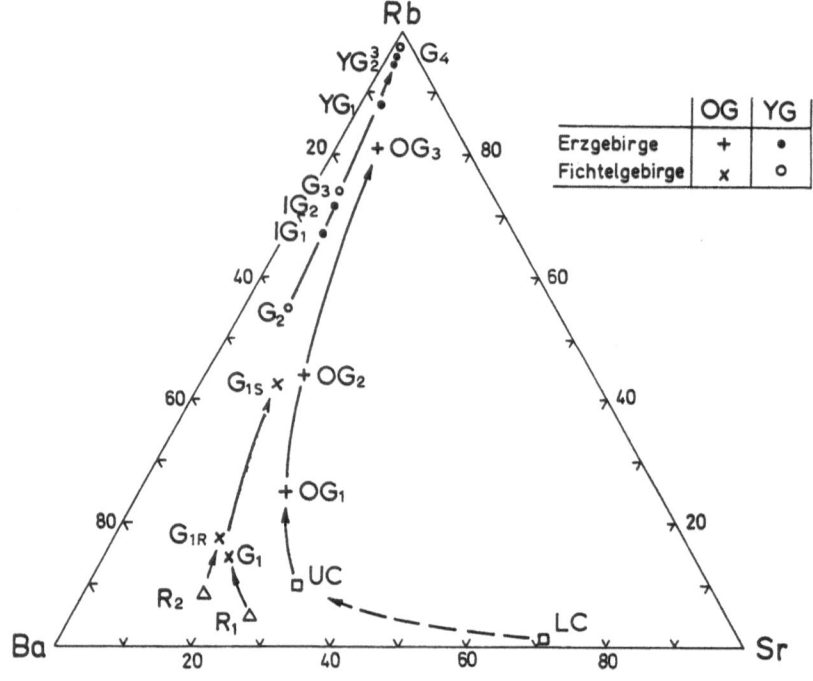

Fig. 24. Compositional trends of the Erzgebirge and Fichtelgebirge granite suites in the Sr-Ba-Rb triangle (Tischendorf et al. 1987:230). The data of the Fichtelgebirge granites are from Richter and Stettner (1979) and represent: R_1 and R_2 marginal facies; G_1, G_{1R}, G_{1S} porphyritic granites of Weißenstadt-Marktleuthen, Reut and Selb; G_2 "Randgranit"; G_3 "Kerngranit"; G_4 "Zinngranit". The YG suite of the Erzgebirge and the G_4 granite of the Fichtelgebirge ("Zinngranit") are hydrothermally overprinted

Sprollenhaus in the Black Forest) a relatively high degree of fractionation (Emmermann 1977). Tin data from Black Forest granites are not available, but mineralogical occurrences of cassiterite on fractures and in greisen bodies in the Sprollenhaus granite suggest a situation close to a tin granite.

3.2 Massif Central, France

The NW part of the French Massif Central hosts over an area of 150 x 50 km several small tin deposits which are associated with Hercynian granite intrusions. Tin mining in the Massif Central dates back to Roman times, but

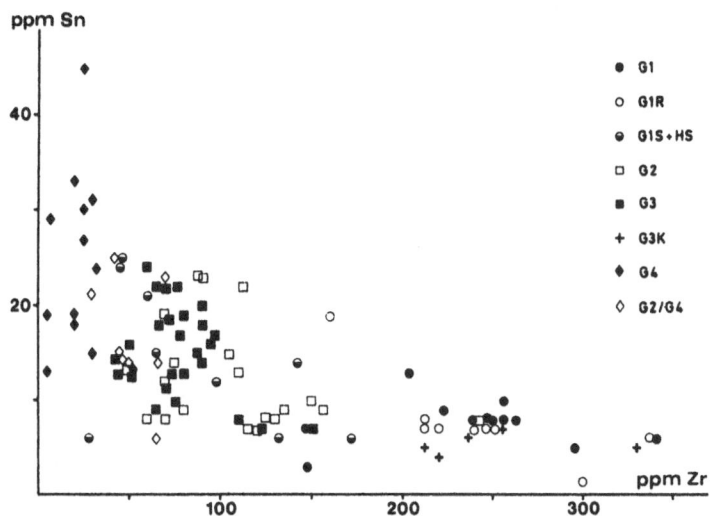

Fig. 25. Sn-Zr variation diagram of the Fichtelgebirge granite suite (Richter and Stettner 1979:110; see explanation of granite types in this reference). The sample group G4 ("Zinngranit") represents the most fractionated granite phase of the Fichtelgebirge and is hydro-thermally overprinted

was, however, never important. There are several geochemical and petrographic studies in relation to exploration work for tin, tungsten and uranium which give whole-rock tin data (Aubert 1969; Ranchin 1970; Burnol 1974, 1978; Boissavy-Vinau 1979; Raimbault 1984).

Tin mineralization is of combined greisen and vein type (Montebras, Échassières, Blond, Saint-Sylvestre) and is bound to locally albitized alkalifeldspar granite stocks which have a Rb-Sr isochron age of 320-300 Ma with Sr initials of 0.707-0.712 (Burnol 1978; Duthou 1978). The high degree of fractionation of these rocks is documented in detail in the above studies. The tin distribution pattern of the granites is given in Fig. 26 as a function of TiO_2 and Rb/Sr. There is a statistically significant linear log-log correlation, in spite of petrographically distinct hydrothermal overprint, which reflects magmatic

Fig. 26 (next page). Tin distribution as a function of TiO_2 and Rb/Sr in granitic rocks of the French Massif Central. Data from Burnol (1978), Boissavy-Vinau (1979), Boissavy-Vinau and Roger (1980). Shale and crust reference compositions from Rösler and Lange (1976), and Taylor and McLennan (1985). The correlation lines are significant at a confidence level of >99.9% (TiO_2-Sn: r=-0.77, n=73; Rb/Sr-Sn: r=0.82, n=55)

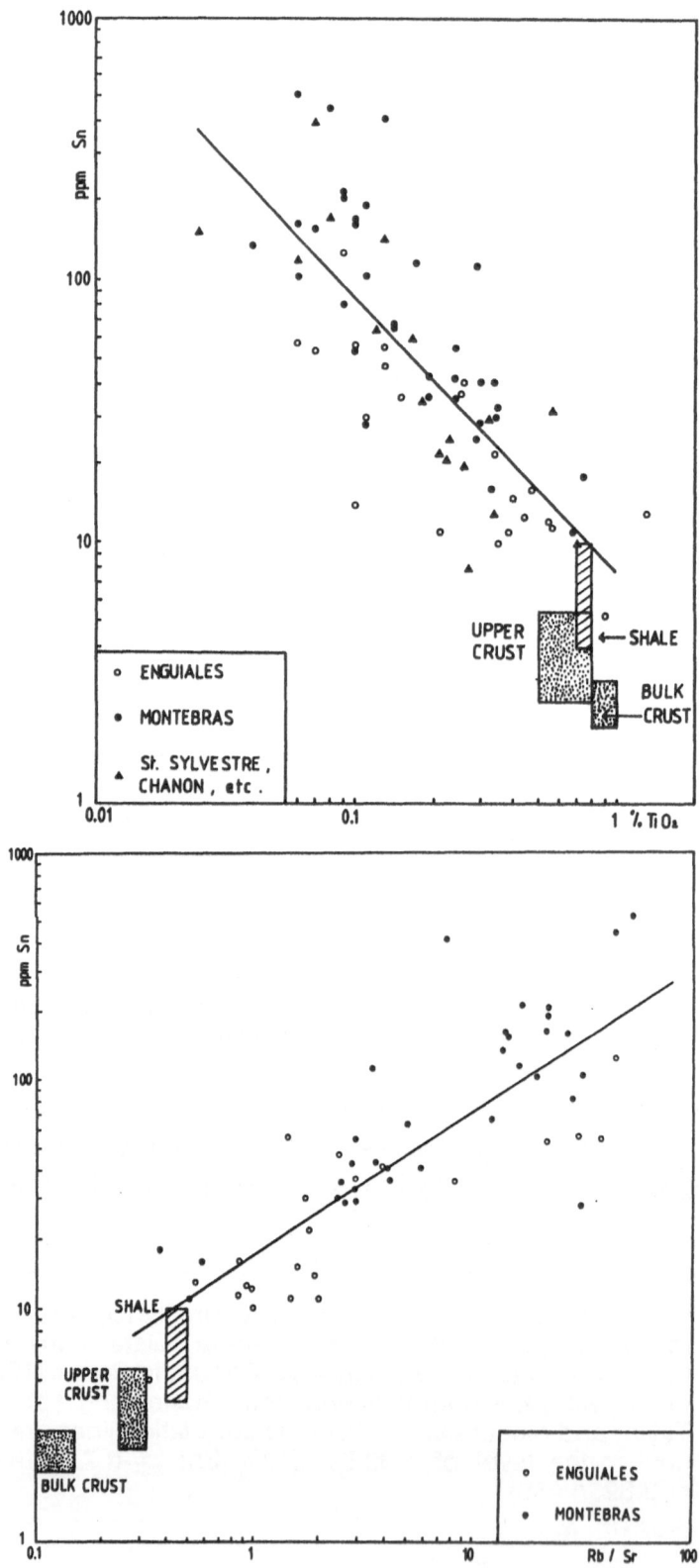

fractionation, in accordance with other trace element data. The fact that this tin enrichment trend reaches up to 100-500 ppm Sn in strongly albitized and muscovitized rock portions indicates a little effective tin redistribution process during hydrothermal overprint, and explains the very limited tin mining potential of this area. The reason for the limited postmagmatic mobility of tin in the Massif Central may lie in the relatively oxidized state of the granites which are situated above the NNO buffer [Raimbault (1984) reports on accessory magnetite]. In contrast, such a situation is favorable for the mobility of uranium which is extensively leached from the granites and which is concentrated in several important ore deposits.

3.3 Cornwall

The Cornwall tin province is, according to both historical and current tin mining figures, the most important European tin producer (mine output in 1988: 3500 t Sn). The hydrothermal Sn-W-As-Fe-Cu-Pb-Zn-Ag-U mineralization in Cornwall is spatially associated with posttectonic Hercynian granites which intrude a 12-km-thick Upper Paleozoic low-grade metasedimentary flysch sequence with subordinate mafic volcanic intercalations (Holder and Leveridge 1986). There are five larger plutons (Dartmoor, Bodmin Moor, St. Austell, Carnmenellis, Land's End) and numerous smaller stocks and dykes which are, according to geophysical data, part of an inferred granite batholith about 250 km by 40 km in size. The total volume of the batholith is estimated at around 68,000 km^3 (Willis-Richards and Jackson 1989).

The granitic plutons have Rb-Sr isochron ages of 280-290 Ma; granite porphyry magmatism extends to 270 Ma (Darbyshire and Shepherd 1985, 1988). Sr and Nd initials ($^{87}Sr/^{86}Sr_i$ 0.709-0.717; ϵ_{Nd} -4.5 to -7.2) correspond to the peraluminous S-type character of the granites and suggest, together with high $\delta^{18}O$ values of 10.8-13.2, an anatectic origin from Proterozoic pelitic material which did not suffer an earlier high-grade metamorphic event (Darbyshire and Shepherd 1985, 1988; Floyd et al. 1983; Jackson et al. 1982). Fluid inclusion Rb-Sr isochron data define a main episode of tin mineralization around 270 Ma (Darbyshire and Shepherd 1985). K-Ar age data from polymetallic veins indicate continued low-temperature hydrothermal activity throughout the Mesozoic and Cenozoic (Halliday 1980; Jackson et al. 1982) which is commonly seen as a consequence of the high total content of heat-producing elements (average values for least altered

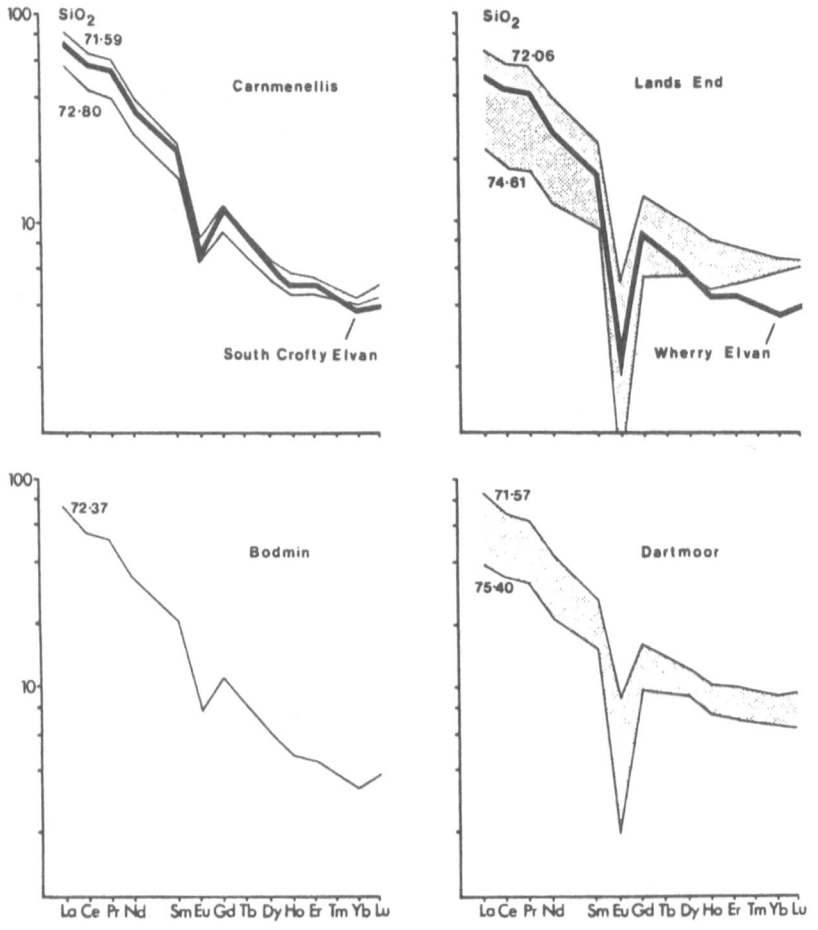

Fig. 27. REE distribution patterns in some granite samples from Cornwall
(Darbyshire and Shepherd 1985:1169)

Cornubian granites are 11.3 ppm U and 19.1 ppm Th; Tammemagi and Smith
1975).

The predominant lithology of the composite batholith is K-feldspar
megacrystic coarse-grained biotite granite (about 90 % of outcrop area). It is
intruded by fine-grained biotite and biotite-muscovite granite, and by
volumetrically subordinate granite porphyry dikes (known as elvans). The
granitic rocks are peraluminous (A/CNK = 1.1-1.4), have low magnetic
susceptibility typical of ilmenite-series granitoids, and have a modal
composition near the 1 kbar thermal minimum of the experimental granite
system. Hydrothermal overprint is very widespread (albitization,
microclinization, muscovitization, tourmalinization, kaolinization; Exley and

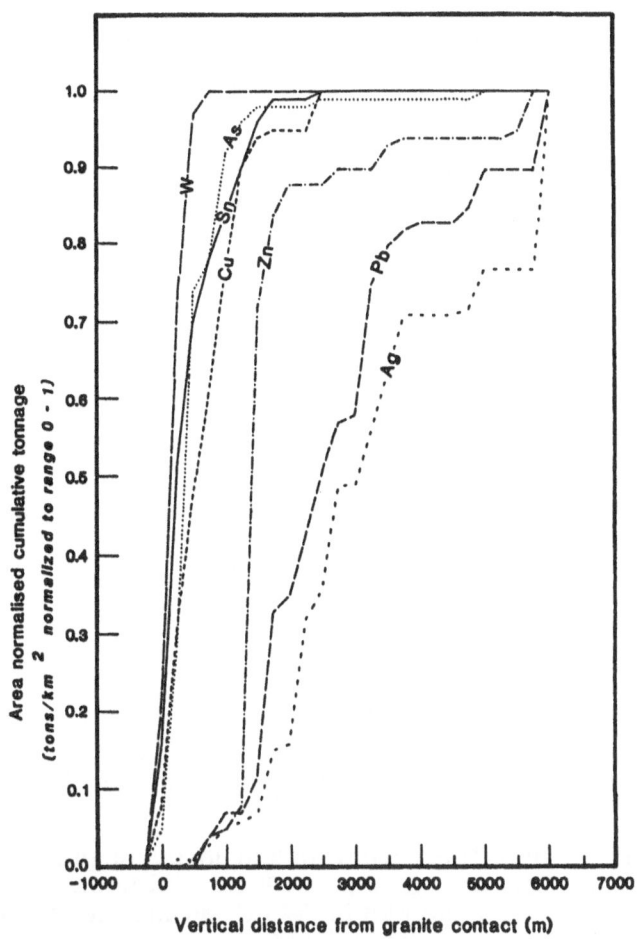

Fig. 28. Metal production of the Cornwall tin province as a function of vertical distance from granite contact. Diagram from Willis-Richards and Jackson (1989)

Stone 1982; Ball and Basham 1984; Charoy 1986). Immobile element patterns indicate pronounced fractionation trends (Ball and Basham 1984; Charoy 1986; Darbyshire and Shepherd 1985). The REE distribution patterns in Fig. 27 for granite phases with minor hydrothermal modification (incipient sericitization and chloritization of feldspars and biotite, respectively) imply strong feldspar fractionation, corroborated by correlation between SiO_2, Rb, Rb/Sr and negative Eu anomaly (Darbyshire and Shepherd 1985). The increasing degree of hydrothermal overprint amplifies the magmatically established REE trends (Alderton et al. 1980).

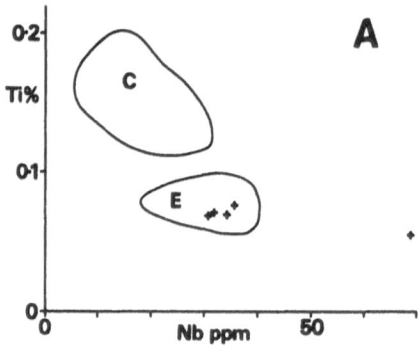

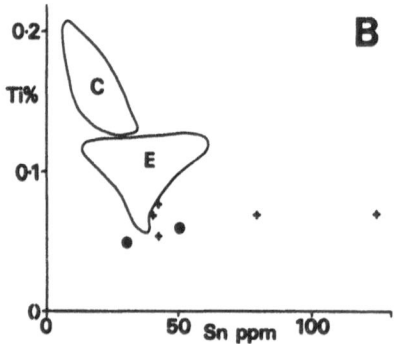

Fig. 29. Ti-Nb (A) and Ti-Sn (B) variation in some granite samples from
Cornwall (Ball and Basham 1984:74). Unpublished data on the
Carnmenellis Granite plot in field C, elvans (quartz porphyry dykes)
from the same area plot in field E. Crosses locate drill samples from
the unexposed Bosworgey Granite, crossed circles mark samples
from the Cligga Head Granite according to Hall (1971)

Tin mineralization is mostly in the form of steeply dipping vein systems and
sheeted vein swarms, and has a strong affinity for the granite contact. Fig. 28
shows the recorded historic mine output for a number of metals in relation to
the vertical distance from the granite contact (Willis-Richards and Jackson
1989). Homogenization temperatures in fluid inclusions give a minimum
temperature range for major cassiterite deposition of the order of 350-450°C
with fluid salinities of 15-23 eq. wt% NaCl (Jackson et al. 1982).

Systematic studies on tin contents in the Cornwall tin granites have not been
published. The data of Stone (1982) and Ball and Basham (1984) hint at tin
enrichment with increasing degree of fractionation (Fig. 29). The highly
evolved nature of the Cornubian granites is graphically summarized in Fig. 30.

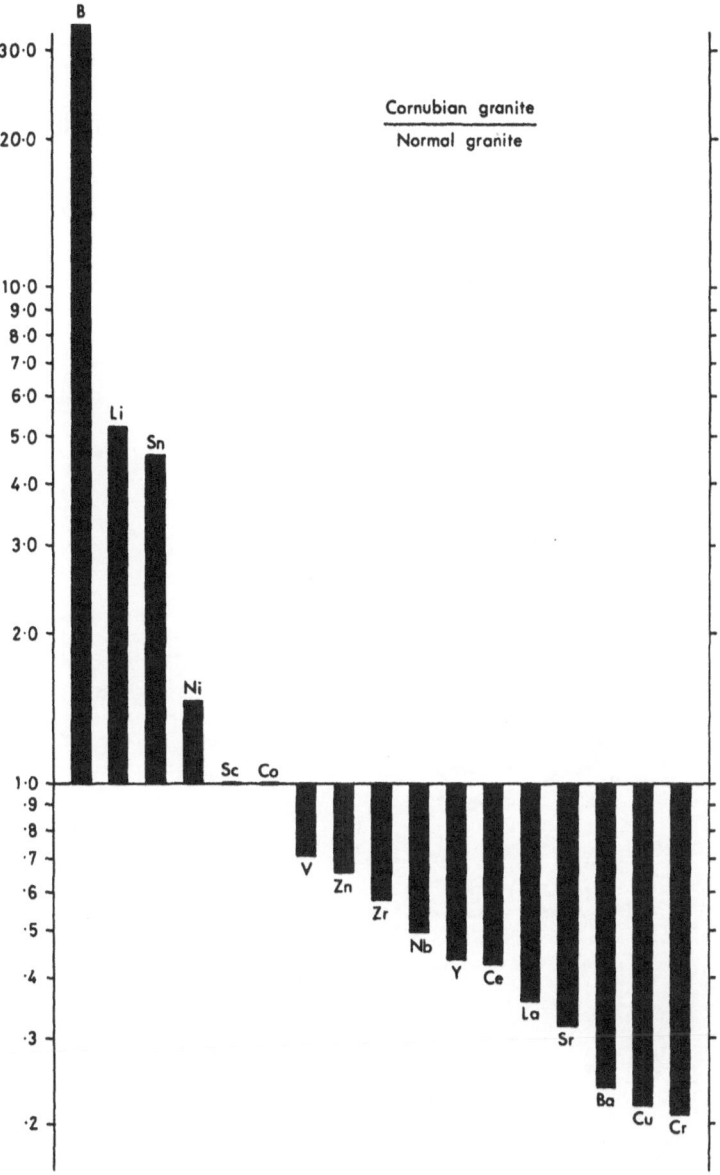

Fig. 30. Trace element characteristics of Cornubian granites (n = 14) normalized to average granite composition as defined by Le Maitre (1976) and Abbey (1983). Diagram from Hall (1990)

The average tin content of the Cornubian granites is given as 14-36 ppm (Stone and Exley 1985), 7-20 ppm (Willis-Richards and Jackson 1989), and 23 ppm (Hall 1990).

66

3.4 Malaysia

The SE Asian tin belt is composed of three petrogenetic-chronologically distinct granite provinces (Fig. 31).

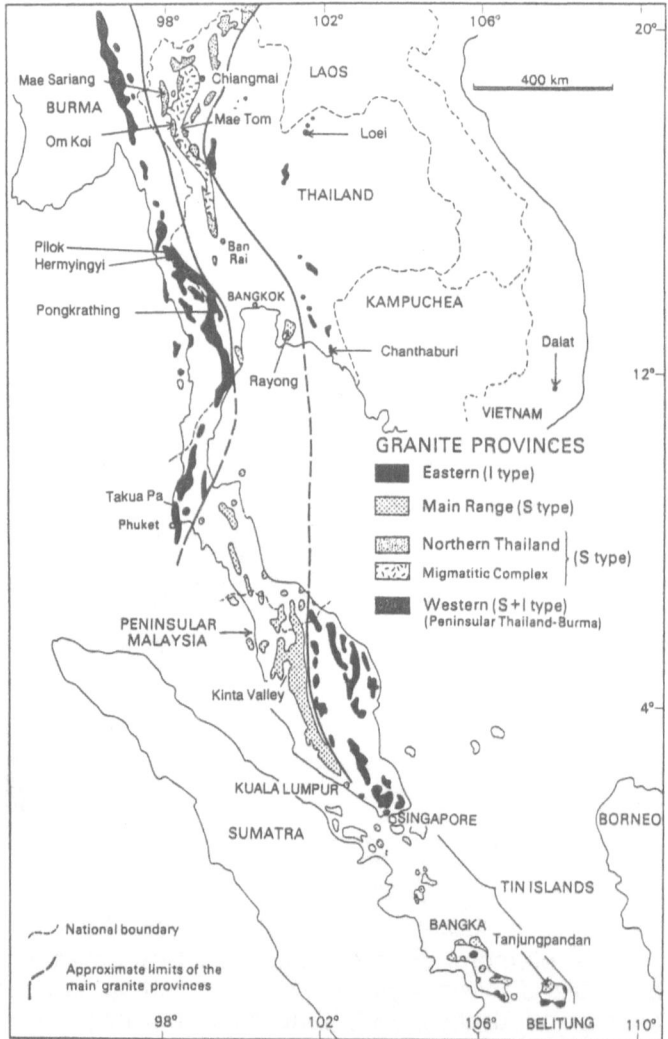

Fig. 31. Geographic distribution of granite provinces in the SE Asian tin belt according to Cobbing et al. (1986), and location of places mentioned in text

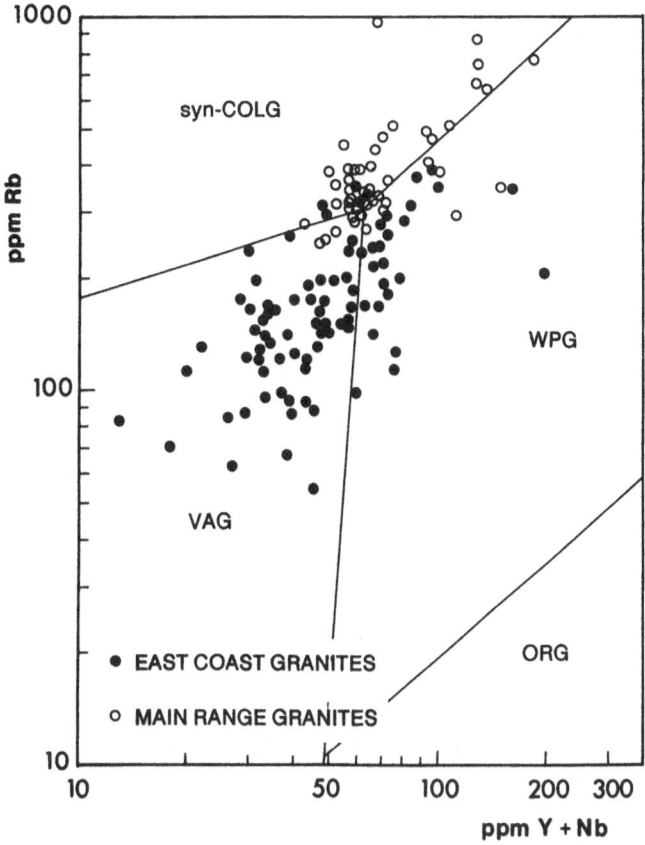

Fig. 32. The composition of the Main Range and eastern province granites of Malaysia in the PEARCE diagram. Data from Cobbing et al. (1986)

1. The Main Range granite province which hosts the famous Malaysian tin fields near Kuala Lumpur and in the Kinta Valley and which is of minor economic importance in central and northern Thailand. The Main Range granites are peraluminous, S-type biotite granites and have Rb-Sr ages in the range 220-200 Ma with Sr initials of 0.716-0.751 (Liew and McCulloch 1985; Darbyshire 1988a). Their geotectonic position is posttectonic with respect to the pre-Permian folding of the Paleozoic country rocks, and is a result of either continental collision of several micro-terranes (Mitchell 1977; Beckinsale et al. 1979) or of intracontinental rifting (Helmcke 1985).

2. The eastern granite province is relatively poor in tin (the ratio of historic tin output of Main Range to eastern granite province in Malaysia is 19:1) and is composed of hornblende-biotite and biotite granites of Permo-Triassic age (265-230 Ma). The granites of the eastern province are chemically more

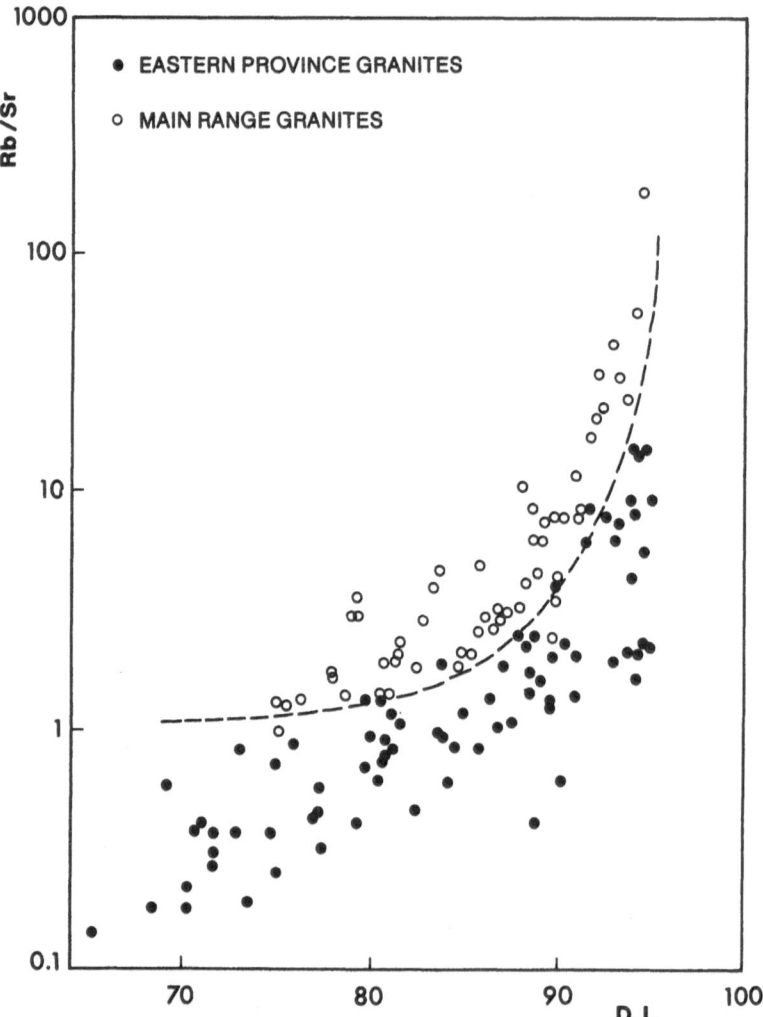

Fig. 33. Rb/Sr as a function of D.I. (Thornton-Tuttle differentiation index) for granitic rocks from Malaysia and southern Thailand: S-type granites of the Main Range province plot different from I-type granites of eastern granite province (influence of mantle material). Data from Pitfield et al. (1987)

primitive compared to the Main Range granites, and classify as volcanic-arc granites in the sense of Pearce et al. (1984) (Figs. 32 and 33). The hornblende-bearing granites have I-type characteristics and their Sr initials are 0.705-0.710, whereas the biotite granites are closer to S-type and have Sr initials of 0.708-0.714 (Liew and McCulloch 1985). The same rock group occurs also in Thailand, where no tin is associated. The prolongation of the eastern granite province into Indonesia is uncertain. The granites of the Tin

Islands appear to represent both petrologically and chronologically a mixed population of Main Range and eastern province (Darbyshire 1988b; Cobbing et al. 1986).

3. The western granite province is restricted to western Thailand and Burma and is composed of Cretaceous-Tertiary granite intrusions in geotectonic relationship to the still active subduction of the Indian Plate below SE Asia. The intrusions consist of metaluminous hornblende-biotite granites (I-type) and peraluminous biotite granites (S-type) with ages of 95-50 Ma and Sr

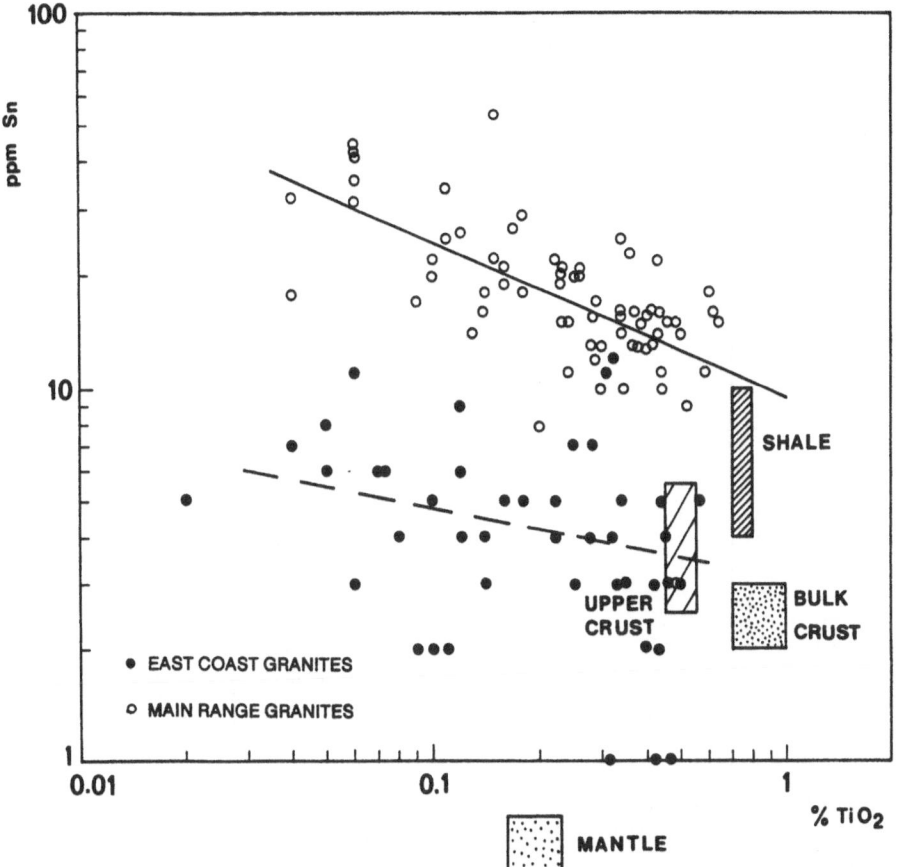

Fig. 34. TiO_2-Sn and Rb/Sr-Sn (see next page) variation diagrams of Main Range and eastern province granites in Malaysia. (Data from Liew (1983); all samples >67 wt% SiO_2). Boxes with reference compositions according to data in Taylor and McLennan (1985) and Rösler and Lange (1976). Correlation for Main Range samples significant with confidence level of >99.9 % (TiO_2-Sn: r=-0.63, n=68; Rb/Sr-Sn: r=0.68, n=70); samples of eastern granite province have log[Rb/Sr]-log[Sn] correlation with 99% cofidence level (r=0.40, n=43), log[TiO_2]-log[Sn] with 80% (r=-0.22, n=43)

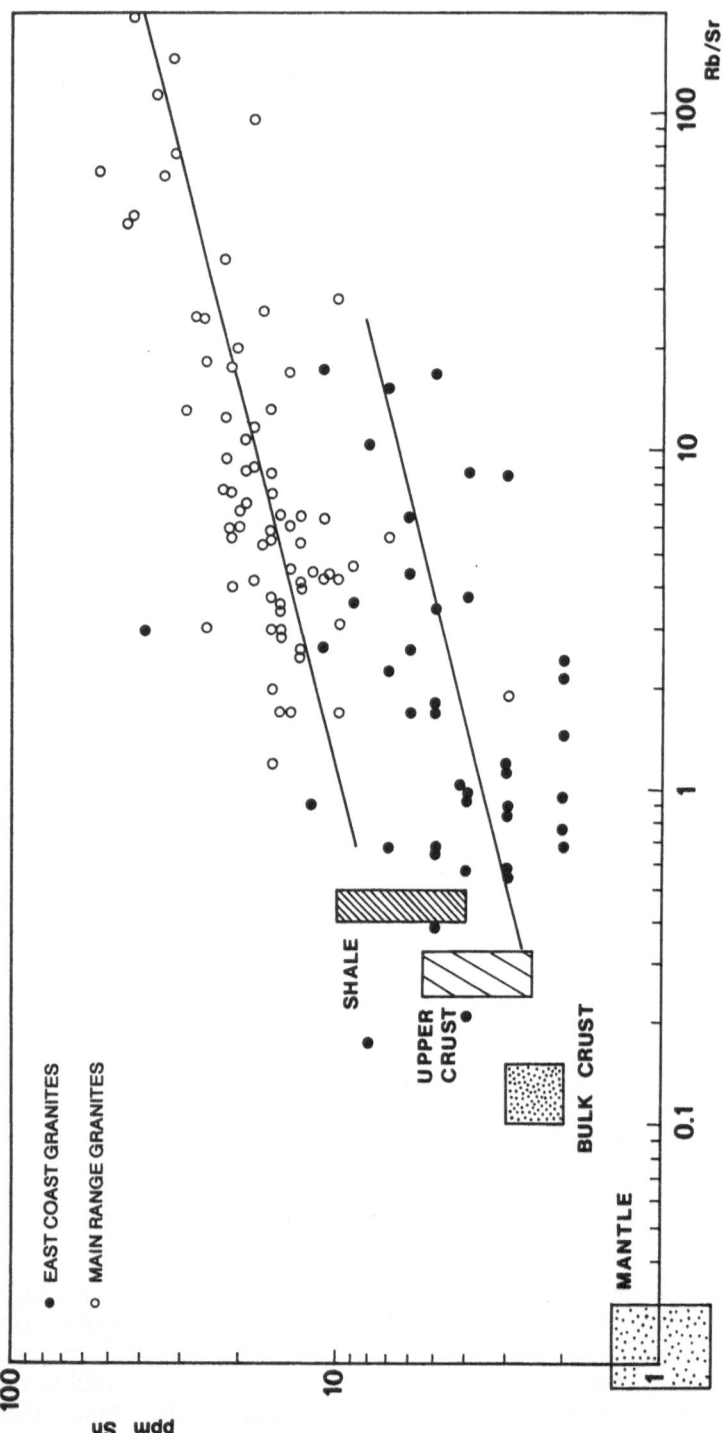

Fig. 34 (continued)

initials of 0.708-0.735 (Beckinsale 1979; Beckinsale et al. 1979; Nakapadung-rat et al. 1984b; Darbyshire 1988c; Darbyshire and Swainbank 1988). The large tin mining areas of Phuket and Phangna in southern Thailand, as well as the Burmese tin deposits, are associated with these Cretaceous-Tertiary biotite granite suites.

The tin distribution pattern of both Main Range and eastern province granites (East Coast) of peninsular Malaysia, based on data in Liew (1983), is compiled in Fig. 34. Both sample populations define statistically significant $\log[Rb/Sr]$-$\log[Sn]$ and $\log[TiO_2]$-$\log[Sn]$ correlation lines, in spite of considerable scatter. Part of this scatter is likely to be a result of hydrothermal overprint, particularly primary dispersion associated with the rich tin mineralization in the Main Range. Another part of this scatter derives from the proximity of tin levels in the eastern granite population near the analytical detection limit. However, the tin contents in both granite populations are distinctly different. Both correlation trends have a similar slope and, at a given Rb/Sr ratio or TiO_2 content, tin content of the Main Range samples is three to four times higher than in those from the East Coast.

The parallel displacement of the two tin enrichment trends suggests different source material and similar magmatic evolution for both rock groups. Their origin is constrained by initial Sr and Nd isotope data (Liew and McCulloch 1985; Darbyshire 1988a). The $\epsilon_{Sr}(T)$-$\epsilon_{Nd}(T)$ diagram in Fig. 35 locates the different compositional fields of both Malaysian granite provinces. The Main Range granites are, in accordance with their mineralogical-geochemical characteristics, of crustal origin. Their source material is probably of Middle Proterozoic age (1700-1500 Ma) as recorded by U-Pb ages of inherited zircons and deduced from Nd model ages (Liew and Page 1985). The East Coast granites, on the other hand, have Sr and Nd initials which suggest the involvement of more primitive source material, in line with their petrochemical characteristics. Nd model ages give a range of 1400-1000 Ma (Liew and McCulloch 1985) and can be interpreted to indicate a mixing process between primitive mantle material and crust (Darbyshire, pers. commun. 1988). It follows from this model that the Main Range tin enrichment trend in Fig. 34 reflects the composition of the metasedimentary basement whereas the East Coast trend is a result of mantle plus basement melting. A linear extrapolation of the TiO_2-Sn variation pattern towards subgranitic composition is not permitted because of the increasingly compatible behaviour of titanium towards mafic composition.

It should be noted that the granite samples plotted in Figs. 32-34 are from the granitic main phases in Malaysia and not from those granite variants directly

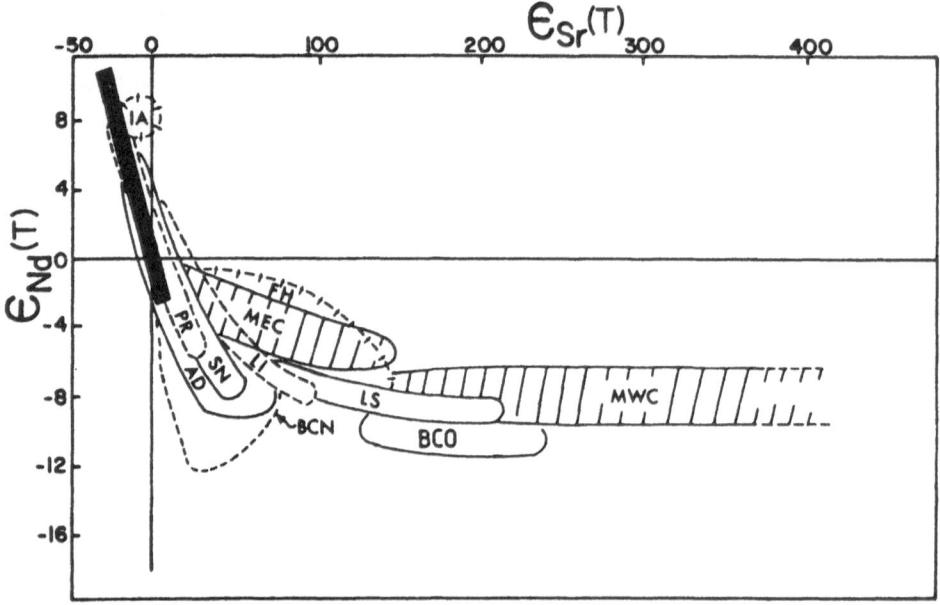

Fig. 35. Initial Sr and Nd isotope composition of Malaysian granites from the Main Range (**MWC**) and eastern granite province (**MEC**) in comparison with other Phanerozoic continental margin granitic rocks. **PR** Peninsular Ranges; **SN** Sierra Nevada; **AD** Central Andean; **LI** Lachlan Foldbelt, I-type; **LS** Lachlan Foldbelt, S-type; **BCO** British Caledonian Older Granites; **BCN** British Caledonian Newer Granites; **FH** French Hercynian, Pyrenees; **black line**: mantle array; **IA** primitive island arc basalts. (Liew and McCulloch 1985:598)

associated with tin mineralization. As in all other tin provinces, the Main Range granites consist dominantly of K-feldspar porphyric medium- to coarse-grained biotite granite which forms large plutons/batholiths. These are locally cut by quantitatively subordinate, medium- to fine-grained subintrusions which are more or less hydrothermally overprinted and which consist of biotite-muscovite to muscovite-tourmaline granite. Tin mineralization is associated with these late muscovite-bearing granite phases.

3.5 Thailand

Tin deposits in Thailand are restricted to the western mountain range near the Burmese border and are associated with both Triassic Main Range granites and, economically more important, Cretaceous intrusions of the western granite province (Fig. 31). Granitic rocks east of Bangkok, such as the Rayong pluton (Main Range type), and all Permo-Triassic intrusions further east, such as the Chanthaburi and Loei granites (eastern granite province) are tin-barren. They are, however, associated with some copper and molybdenum mineralization of porphyry type (Brown et al. 1951; Jacobson et al. 1969; Lehmann 1988a).

A regional petrographic-geochemical study in central and northern Thailand and in the Hermyingyi Mine (Burma) compared the following granite intrusions, located in Fig. 31:
1. Mae Sariang pluton (Triassic): K-feldspar megacrystic biotite-hornblende granite; metaluminous to weakly peraluminous (I-type); no tin mineralization.
2. Om Koi pluton (Triassic): K-feldspar megacrystic biotite and biotite-muscovite granite (sub-solidus muscovite); peraluminous (S-type); minor tin mineralization in associated pegmatite and hydrothermal systems (quartz-tourmaline-cassiterite-wolframite-muscovite-biotite-kaolin veins).
3. Mae Tom pluton (Cretaceous): K-feldspar megacrystic biotite granite; weakly peraluminous (I-type); no tin mineralization.
4. Loei intrusions (Permo-Carboniferous): a granitic suite with a wide range of compositions from hornblende quartz monzonites to hornblende-biotite granodiorites to biotite-hornblende granodiorites/granites and to biotite granites; metaluminous to weakly peraluminous (I-type); no tin, but copper porphyry mineralization.
5. Chanthaburi intrusions (Permo-Carboniferous): a suite of K-feldspar megacrystic biotite-hornblende granites to biotite granites; metaluminous to weakly peraluminous (I-type); no tin, but molybdenum porphyry mineralization.
6. Rayong pluton (Triassic): K-feldspar megacrystic biotite and biotite-muscovite granite (sub-solidus muscovite); peraluminous (S-type); $^{87}Sr/^{86}Sr_i$ 0.726 (Nakapadungrat et al. 1984b); no tin mineralization.
7. Border Range granites (Pongkrathing, Pilok, Hermyingyi) (Cretaceous and early Tertiary): K-feldspar megacrystic biotite and biotite-muscovite granite (sub-solidus muscovite) with alkalifeldspar aplogranite subintrusions; peraluminous (S-type); Hermyingyi: $^{87}Sr/^{86}Sr_i$ 0.727 (Lehmann and Mahawat 1989); tin mineralization of greisen, stockwork, vein and sulphide-replacement type associated chiefly with aplogranites and aplite-pegmatite systems.

All these granitic intrusions have a high emplacement level and are partly (eastern granite province) or completely (Main Range and western granite province) equilibrated with 1 ± 0.5 kbar minimum-melt conditions in the

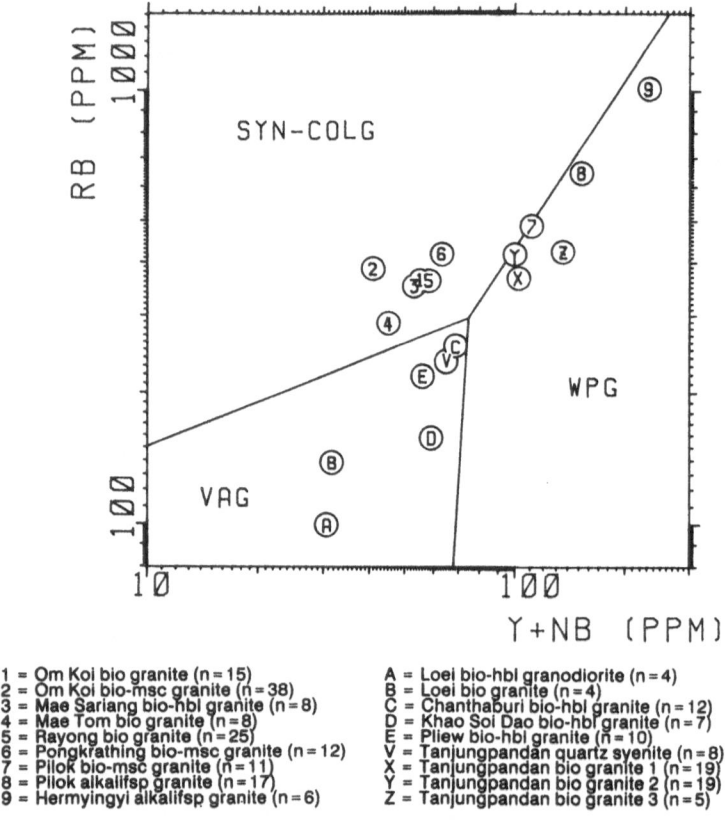

1 = Om Koi bio granite (n = 15)
2 = Om Koi bio-msc granite (n = 38)
3 = Mae Sariang bio-hbl granite (n = 8)
4 = Mae Tom bio granite (n = 8)
5 = Rayong bio granite (n = 25)
6 = Pongkrathing bio-msc granite (n = 12)
7 = Pilok bio-msc granite (n = 11)
8 = Pilok alkalifsp granite (n = 17)
9 = Hermyingyi alkalifsp granite (n = 6)

A = Loei bio-hbl granodiorite (n = 4)
B = Loei bio granite (n = 4)
C = Chanthaburi bio-hbl granite (n = 12)
D = Khao Soi Dao bio-hbl granite (n = 7)
E = Pliew bio-hbl granite (n = 10)
V = Tanjungpandan quartz syenite (n = 8)
X = Tanjungpandan bio granite 1 (n = 19)
Y = Tanjungpandan bio granite 2 (n = 19)
Z = Tanjungpandan bio granite 3 (n = 5)

Fig. 36. The composition (arithmetic means) of granite populations from Thailand, Burma and Indonesia (Tanjungpandan Pluton, Belitung Island) in the Pearce diagram. SYN-COLG syn-collisional granites, VAG volcanic-arc granites, WPG within-plate granites, according to the terminology of Pearce et al. 1984)

experimental Qz-Ab-Or-An-H_2O system. Systematic trace-element trends imply for all granite populations fractional crystallization as the dominant process controlling magmatic evolution. The tin-bearing alkalifeldspar aplogranites display an extreme degree of differentiation which has no petrological equivalent in the eastern granite province (Lehmann and Mahawat 1989).

Analogous to the situation in Malaysia, the granites of the Main Range and western province are located predominantly in the "syn-collision" reference field of the Pearce diagram, i.e. crustal source material, whereas the more primitive granitic rocks of the eastern province plot in the "volcanic-arc" reference field (Fig. 36). Included in Fig. 36 are the four main intrusive phases of the Tanjungpandan pluton from Belitung Island, Indonesia, which are discussed later (Chap. 4.1). All granite populations in this figure are of posttectonic position with respect to the early Permian regional folding event, and are not foliated. The trend of increasing Rb and Y+Nb contents in the most evolved granite phases (along the "syn-collision"-"within-plate" division line in Fig. 36) is a result of intramagmatic fractionation and not of changing source rock composition as usually implied in the petrogenetic interpretation of such diagrams (Pearce et al. 1984).

The tin distribution pattern for these granite populations is given in Fig. 37 which compares average tin contents (arithmetic means ± one standard deviation) with degree of differentiation. The parameters used as indicators of differentiation are TiO_2, Rb/Sr and D.I. (normative Qz+Or+Ab, i.e. Thornton-Tuttle differentiation index). Tin mineralization is associated only with those

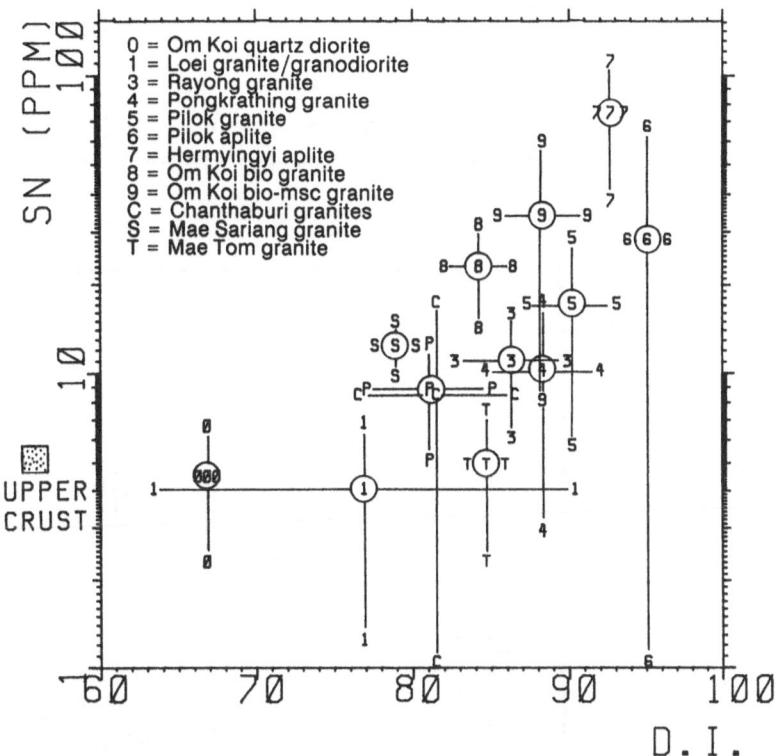

Fig. 37. Tin content as a function of D.I. (normative Qz+Ab+Or, i.e. Thornton-Tuttle differentiation index), and of TiO_2 (wt%) and Rb/Sr (see next page) for several granite populations from Thailand and Burma. Important tin mineralization is associated with Hermyingyi and Pilok aplogranites. Global reference fields from Taylor and McLennan (1985)

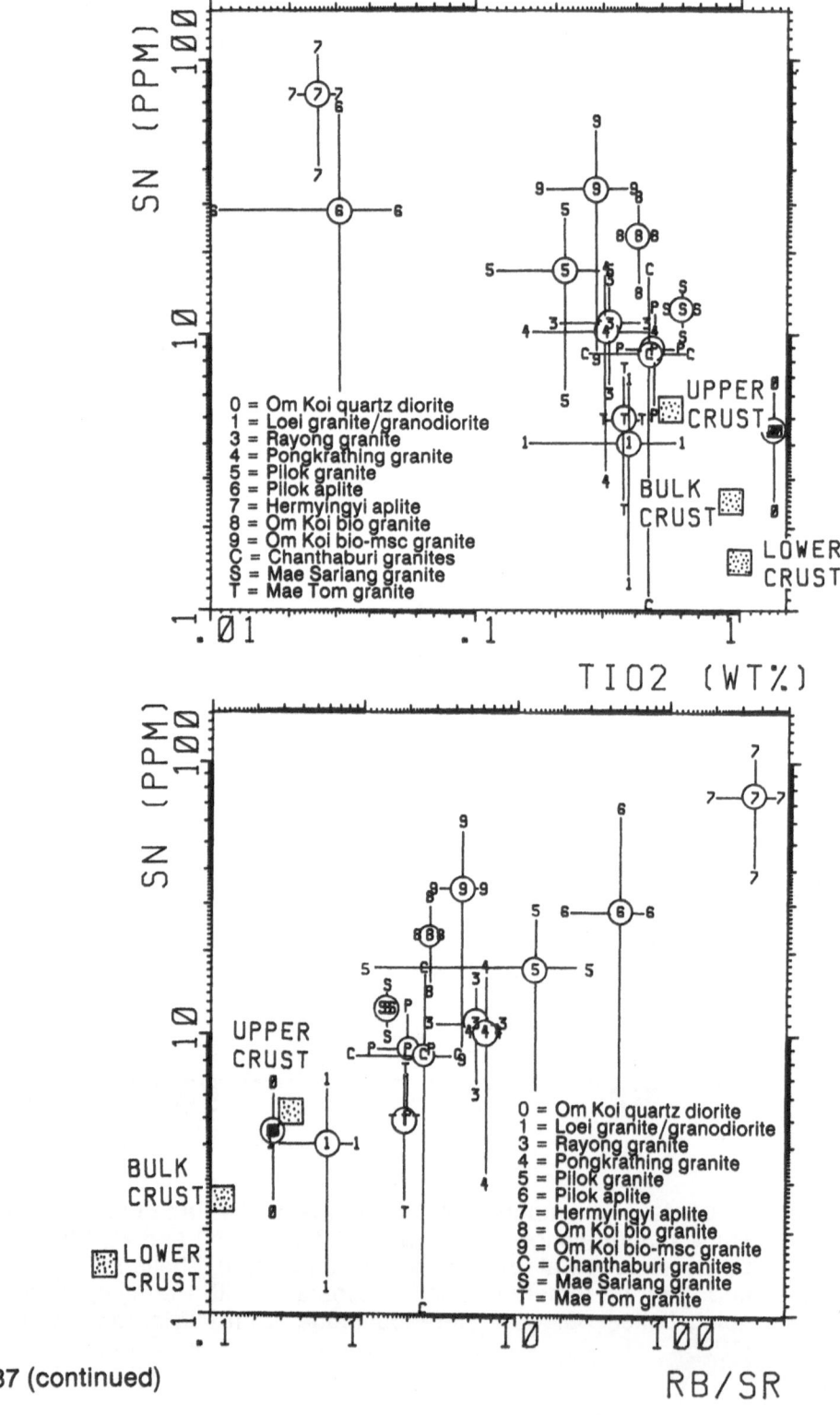

Fig. 37 (continued)

rocks which have the most differentiated composition. These rocks plot on the most evolved part of a general tin enrichment trend which extrapolates back to average crustal composition. The highly fractionated Pilok aplogranite has a very large scatter in tin content which is a result of hydrothermal tin redistribution and which will be discussed in Chapter 4.

3.6 Nigeria

About ninety percent of the Nigerian tin production comes from placer deposits associated with greisens, albitization zones and vein swarms in apical portions of Jurassic biotite granite intrusions (Buchanan et al. 1971; MacLeod et al. 1971; Bowden and Kinnaird 1984). There is also a small amount of tin and tantalum produced from deeply weathered pegmatites of Middle-Cambrian age in a 400-km-long, SW-NE-aligned belt between Ife and Jos (Matheis and Caen-Vachette 1983).

The Mesozoic biotite granites are part of a petrologically extended ring-complex suite and occur locally along a more than 1000-km-large lineament zone which stretches from the Air Plateau in Niger in the north to the Jos Plateau in central Nigeria to the south. The ring complexes have dimensions in the 1- to 10-km range and consist of an often eroded superstructure of trachytic to rhyolitic rocks intruded by high-level alkalifeldspar-biotite granites and quartz syenites of metaluminous to peraluminous composition with variable mineralogy (hornblende, riebeckite, hedenbergite, fayalite, etc.). The age of this anorogenic magmatism decreases systematically towards the south, with an age of 164 ± 4 Ma in the Jos Plateau. This trend has been interpreted as related to a stationary thermal anomaly in the mantle (Sillitoe 1974; Breemen et al. 1975).

The ring complexes consist of composite A-type intrusions (Collins et al. 1982) which have trace element distribution patterns typical of fractional crystallization suites (diagnostic elements are particularly the least mobile elements Zr, Ti, Nb, Y). Tin mineralization is associated with most fractionated and peraluminous granite phases (Imeokparia 1980, 1984, 1986a,b; Olade 1980). Sr initials of biotite granites with little hydrothermal overprint are in the range 0.706-0.709 and indicate an origin from the lower crust or from mantle with crustal contamination. Most fractionated granite phases with hydrothermal overprint have Sr levels of a few ppm only (sensitivity against Sr

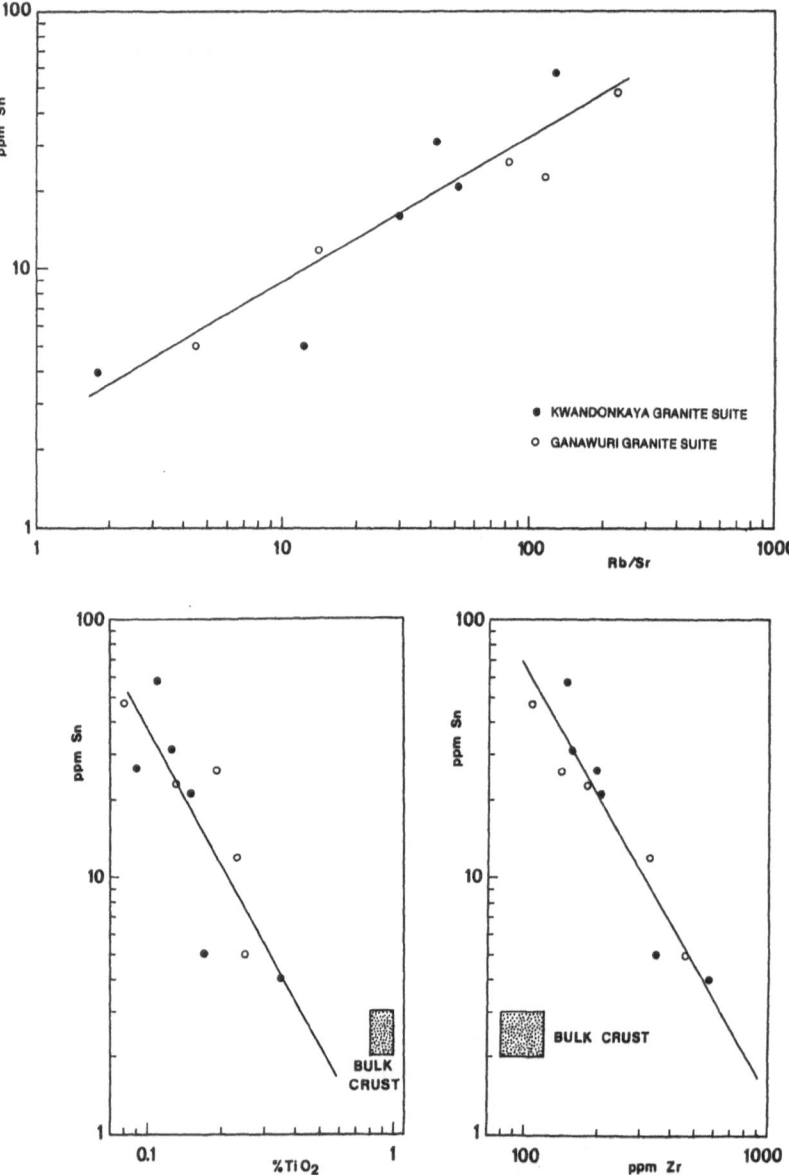

Fig. 38. Tin content as a function of Rb/Sr, TiO$_2$ (wt%) and Zr (ppm) in the Jurassic granite suites of the Kwandonkaya and Ganawuri ring complexes in central Nigeria (in both suites: hornblende-fayalite granite to biotite microgranite). The data are arithmetic means of individual granite units and represent 38 samples from Ganawuri and 108 samples from Kwandonkaya (Imeokparia 1984, 1986a). All correlation lines are statistically significant at a confidence level of >99% (log[TiO$_2$]-log[Sn]: r=-0.83, n=11; log[Rb/Sr]-log[Sn]: r=0.92, n=11; log[Zr]-log[Sn]: r=-0.95, n=11)

exchange) and record heterogeneous $^{87}Sr/^{86}Sr_i$ values up to 0.752 (Breemen et al. 1975).

The tin distribution patterns for two tin-bearing granite suites from the central part of the Jos Plateau are shown in Fig. 38, based on data in Imeokparia (1984, 1986a). The linear correlation lines are in accordance with a magmatic evolution controlled by fractional crystallization; tin contents of least evolved granite samples are near Clarke values. The relatively high Zr contents result from the alkali-rich melt and high melt temperature (hypersolvus composition) which allow high zircon solubility (Bowden 1982; Watson and Harrison 1983).

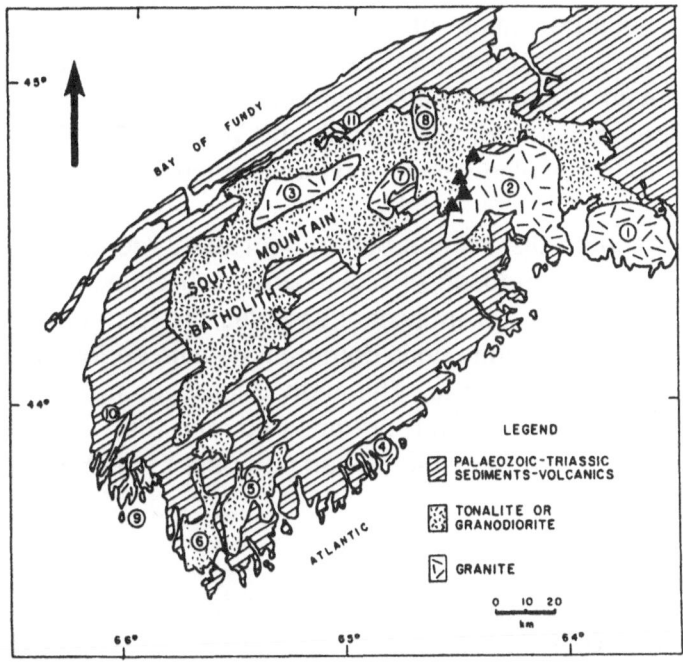

Fig. 39. Geological outline of SW Nova Scotia and distribution of the Halifax (1), New Ross (2), and West Dalhousie (3) plutons mentioned in text. Tin prospects are located by solid triangles. (According to Smith and Turek 1976, and Smith et al. 1982)

3.7 Nova Scotia, Canada

A large part of the Nova Scotia peninsula in eastern Canada is composed of Devonian granitic rocks which intrude a 12-km-thick, Lower Paleozoic volcano-sedimentary sequence in low-grade metamorphic facies (Fig. 39).

The largest intrusion in Nova Scotia is the South Mountain batholith, which is exposed over 6000 km^2. It consists of several little-mapped subintrusions with gradual contacts from peripheral biotite granodiorite towards biotite and biotite-muscovite granite in the central parts. The petrological and chemical zonation is explained by a process of in situ differentiation (Smith 1979; Smith and Turek 1976). There are similarities to the Blue Tier batholith in Tasmania, Australia (Groves and McCarthy 1978), which has, however, much more tin. Tin mineralization in the South Mountain batholith is restricted to several tin prospects in the New Ross pluton (Fig. 39).

The zoned plutons of the South Mountain batholith are 380-350 Ma old and have initial Sr isotope ratios of 0.708 ± 3. They have peraluminous composition (accessory cordierite and andalusite; 2-4 wt% normative corundum), contain abundant metasedimentary xenoliths, and satisfy the criteria for S-type granitic rocks (Chappell and White 1974). Their magmatic evolution is controlled by plagioclase and biotite fractionation (Smith 1979).

Figure 40 shows the tin distribution patterns for the three largest subplutons which are dominantly composed of biotite granite (Halifax, New Ross, West Dalhousie). Chemical data for individual samples are not published, but the arithmetic means for ten granite units demonstrate a tin enrichment trend very similar to that in the Erzgebirge. The high degree of fractionation of the Erzgebirge tin granites is, however, not attained, and only few samples from the weakly tin-bearing New Ross pluton can compare to the Younger Granites in the Erzgebirge.

3.8 Cape Granite, South Africa

The post- or anorogenic Cape batholith consists of several smaller granite plutons which form a 200-km-long, NW-SE-orientated belt near Capetown. The granites are around 600 Ma old and represent a differentiation suite with high intrusion level which goes from K-feldspar megacrystic coarse-grained

---

Fig. 40 (next page). Tin content of three plutons from the South Mountain batholith in Nova Scotia as a function of TiO$_2$ (wt%) and Rb/Sr. Data points are arithmetic means (horizontal and vertical bars for Halifax samples indicate one standard deviation) and are from Smith (1979) and Smith et al. (1982)

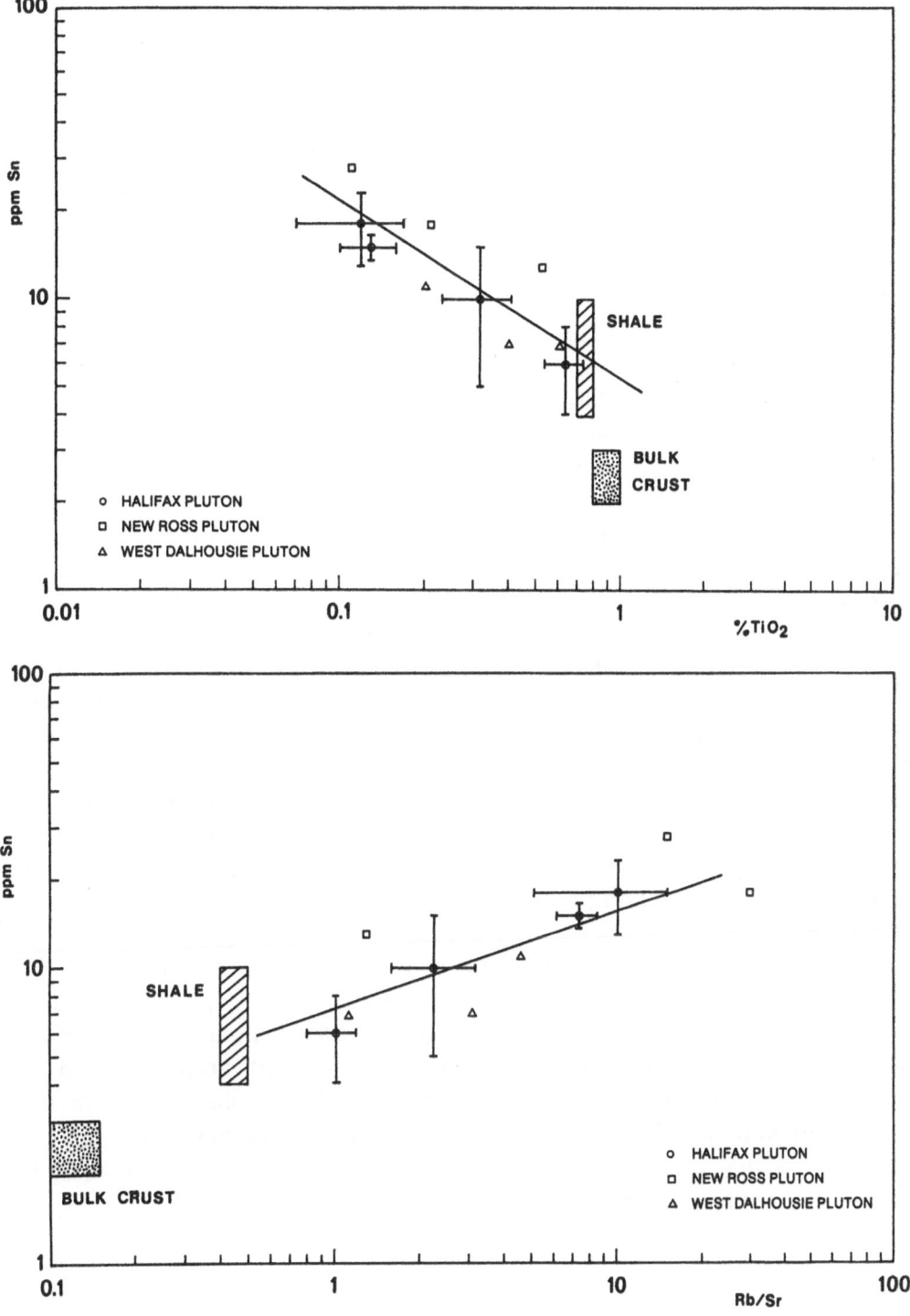

Fig. 40. For legend see previous page

biotite granite (main phase) to medium-grained biotite granite and to fine-grained biotite-muscovite and alkalifeldspar granite in peripheral subunits (Kolbe 1966). Hornblende, cordierite and titanite are accessory components in the coarse-grained granite phase, tourmaline is abundant in the medium- and fine-grained phases. There are several prospects with quartz-tourmaline-muscovite-cassiterite-pyrite-arsenopyrite veins in the endo- and exocontact of the granites (Malmesbury shale-quartzite sequence), and small tin placer deposits were sporadically mined prior to World War II (Thamm 1943; Hunter 1973). Disseminated molybdenite and breccia pipes with pyrite-molybdenite-scheelite in fluorite-bearing alkalifeldspar granite have been described by Scheepers and Schoch (1988).

Based on geochemical data, Kolbe (1966) and Kolbe and Taylor (1966b) interpreted the Cape granites as a comagmatic fractionation suite. The porphyritic main phase has a composition near average "low-Ca granite" (Turekian and Wedepohl 1961), the two younger granite phases display an advanced degree of fractional crystallization and are modified by fluid interaction and resultant loss of mobile elements (Kolbe 1966). Low normative corundum content (<1 wt%) and metaluminous to weakly peraluminous composition (mol. Al_2O_3/Na_2O+K_2O+CaO <1.1) in all granite phases point to an I-type origin; high Nb, Ga and Y contents in the alkalifeldspar granites are similar to those of anorogenic, within-plate granites (A-type). Biotite compositions from leucogranites indicate oxygen fugacities between the Ni-NiO and hematite-magnetite buffers (Scheepers and Schoch 1988), and Fe_2O_3/FeO rock ratios of ≥0.5 are indicative of a magnetite-series affiliation. Pervasive hydrothermal alteration under oxidizing conditions is accompanied by molybdenum and uranium redistribution (Th/U 10-20; Scheepers and Schoch 1988; Schoch and Scheepers 1990).

The tin distribution pattern of the Cape granite suite is shown in Fig. 41. The behaviour of tin is distinctly different from elements like Cs, Rb, Sr, Co, Ni, V and Ti, which display systematic enrichment and depletion trends (Kolbe 1966). The samples have a nearly constant tin content around 3 ppm and there is no significant dependence on degree of fractionation, i.e. coarse-grained main phase with 3 ppm Sn, medium-grained phase with 3.2 ppm Sn, and fine-grained phase with 3.4 ppm Sn. A similar situation is seen in the Snowy Mountains granites of SE Australia.

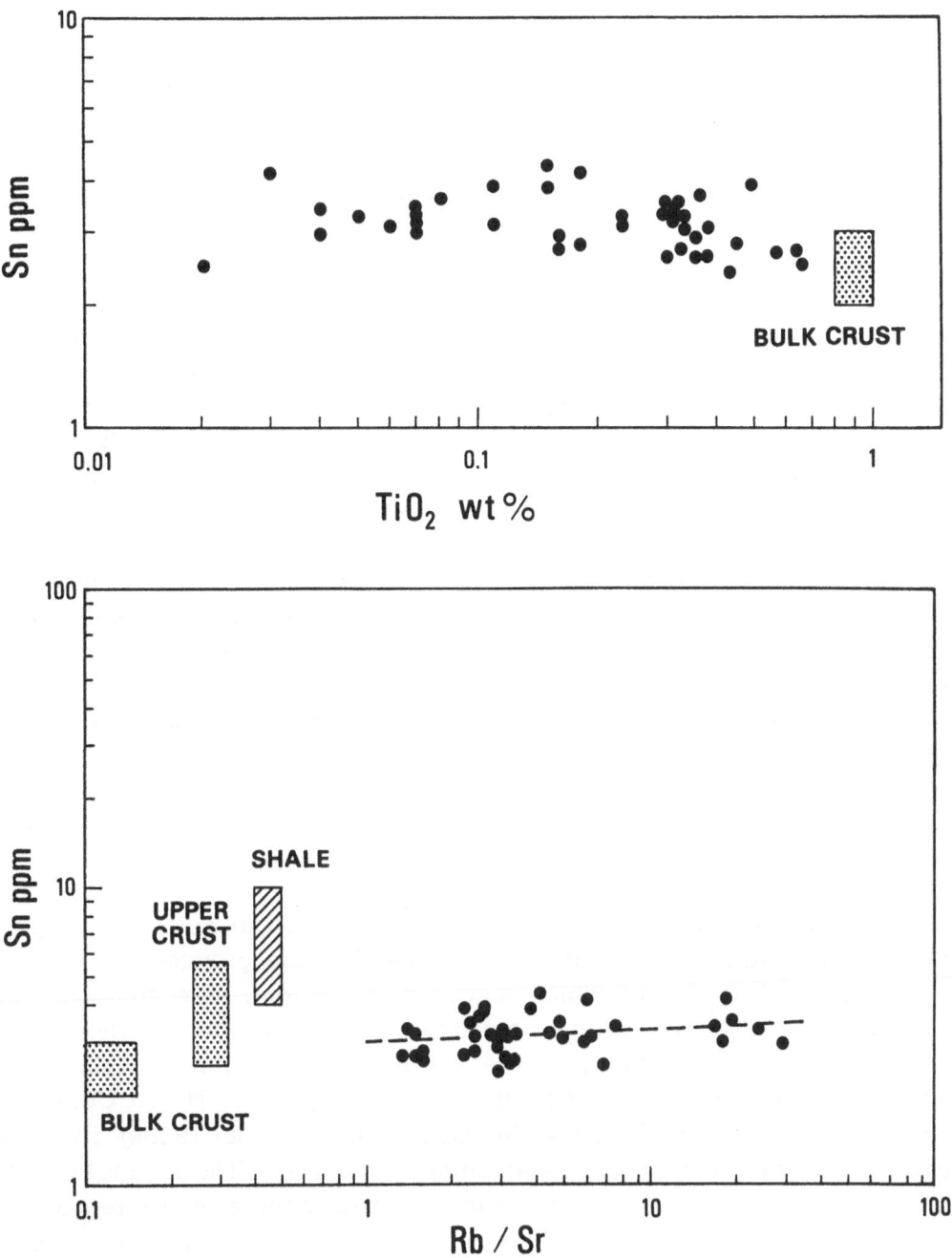

Fig. 41. Tin content as a function of TiO$_2$ (wt%) and Rb/Sr in the Cape granite suite, South Africa. (Data from Kolbe 1966)

3.9 Snowy Mountains, SE Australia

The approximately 200 x 100-km-large region of the Snowy Mountains in New South Wales, SW of Canberra, is part of the Lachlan Foldbelt and consists chiefly of Silurian granitic rocks intruded into low-grade clastic sediments (flysch) of Ordovician age. The granitic rocks are locally foliated, have discordant contacts and high intrusion level, and form a composite intrusion suite of tonalitic to leucogranitic composition. The chemical variation of the mainly granodioritic plutons has been interpreted by White and Chappell (1977, 1983) as a result of mixing of anatectic partial melts and of restitic material. The leucogranitic late phases with a composition near the low-pressure thermal minimum of the experimental $Qz-Ab-Or-H_2O$ system are characterized by an advanced degree of fractional crystallization (Kolbe and Taylor 1966a,b) and have a less peraluminous (locally even metaluminous) composition compared to the granodioritic main phase. The leucogranites satisfy some I-type criteria, the biotite granodiorites are of S-type (White and Chappell 1977, 1983; Hine et al. 1978). Initial Sr and Nd isotope data of the S-type granitic rocks are in the range of 0.709-0.718 and ϵ_{Nd} -6 to -10, respectively, and suggest a crustal origin from 1400 Ma old basement (Nd model age). For the I-type leucogranites, on the other hand, values of $^{87}Sr/^{86}Sr_i$ 0.705-0.712 and $\epsilon_{Nd}(T)$ 0 to -9 document involvement of mantle material (Compston and Chappell 1979; McCulloch and Chappell 1982). The leucogranites of the Lachlan Foldbelt are classified as magnetite-series rocks (White and Chappell 1983).

Tin data for the Snowy Mountains granites are given in Kolbe and Taylor (1966a) and are plotted in Fig. 42. The samples are grouped into biotite granodiorite (S-type) and biotite leucogranite (I-type). The granodiorites occasionally have accessory cordierite and muscovite, rarely green hornblende, and correspond to average high-Ca granites (Turekian and Wedepohl 1961). The leucogranites have low Ca, Fe, Mg, Cr, Ni, Co, Cu, V, Zr, and Sr contents and are high in U, Rb, Cs (Kolbe and Taylor 1966a). Their Sn contents are, however, constant and range from 2-4 ppm. The absence of any tin enrichment in these rocks in spite of a high degree of differentiation, similar to the case of the Cape granite, is probably understandable as a consequence of conditions of high oxygen fugacity (magnetite-series granites) and of a low degree of alumina saturation (metaluminous to weakly peraluminous composition).

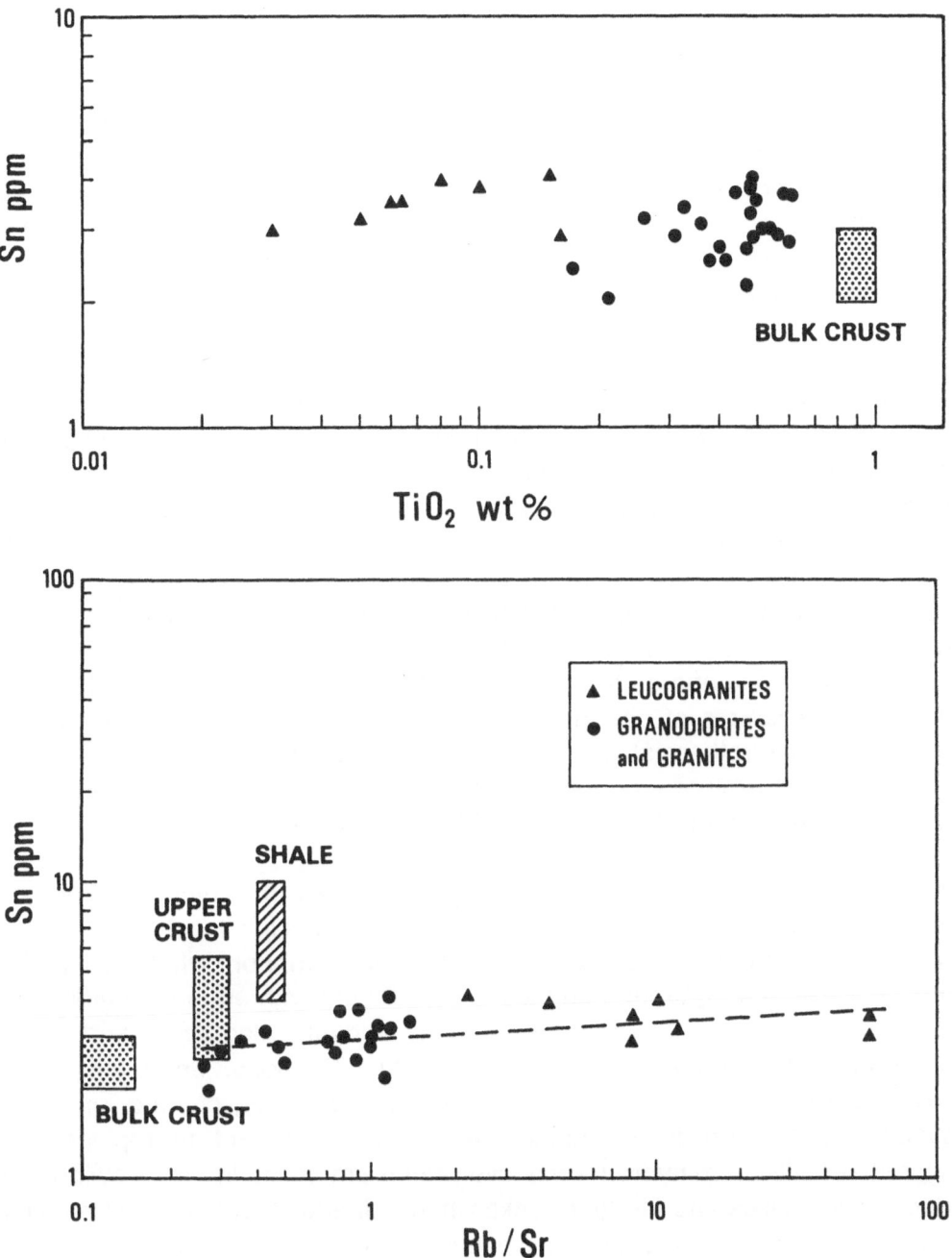

Fig. 42. Tin contents as a function of TiO$_2$ (wt%) and Rb/Sr in granitic rocks from the Snowy Mountains, Lachlan Foldbelt, Australia. Data from Kolbe and Taylor (1966a)

4 Hydrothermal Redistribution

Shallow igneous intrusions produce thermal perturbations in fluid-rich upper crustal rocks which inevitably generate hydrothermal systems in which thermal energy is dispersed by a combination of convective fluid circulation and heat conduction (Norton and Knight 1977). Permeability is the fundamental rock property which controls convective fluid motion and associated chemical and mineral alteration (Norton and Knapp 1977; Norton 1988).

Stable isotope patterns in and around epizonal magmatic intrusions prove that hydrothermal systems can reach many kilometres beyond the magmatic contact zones, with depths of penetration up to 10 km (Taylor 1977, 1979; Criss and Taylor 1986). Important parts of such systems are often affected by little or no petrographically appreciable mineral alteration. High-temperature fluid circulation at >450 $^{\circ}$C can leave original igneous textures and mineral assemblages in granitic rocks essentially preserved because the assemblage biotite-plagioclase-K-feldspar-quartz is compatible with a broad range of fluid compositions at temperatures even lower than 450 $^{\circ}$C (Helgeson et al. 1978). A classical example of dominantly high-temperature fluid circulation is the Skaergaard hydrothermal system in which plagioclase (An_{69} to An_{34}) is stable, and only minor amounts of alteration minerals are formed during the waning stages of hydrothermal activity (Norton and Taylor 1979).

In granitic rocks in general and in tin granite suites in particular, abundant secondary fluid inclusions along sealed microfractures in all rock components (including plagioclase) are a widespread phenomenon which points to recrystallization of igneous minerals in equilibrium with aqueous fluids. The same rocks show, however, also to a variable degree - and often predominantly - features of non-equilibrium conditions, i.e. blastesis of secondary hydrous minerals such as sericite/muscovite, chlorite, prehnite, clinozoisite etc. on intergranular spaces, fractures and in replacement aggregates. The term hydrothermal overprint is employed in the following text also in such cases and for such rocks where mineral alteration is not a major feature, but where the geochemical element distribution pattern points to important postmagmatic-hydrothermal fluid interaction and chemical alteration.

Fluid flow in pluton-host rock environments occurs predominantly through percolation networks on all scales (grain boundaries, micro- and macrofractures). The magnitude of fluid flow is proportional to rock permeability, which has a tendency to decrease during the process of cooling and concomitant mineral precipitation (eventually hydrothermal mineralization). This is a consequence of the generally positive correlation between temperature and solubility for most minerals. Permeability can be drastically increased during the early history of a magmatic-hydrothermal system by sequential exsolution of a magmatic fluid phase which is of fundamental importance in porphyry-type (sensu lato) ore environments.

A characteristic feature of many mineralized granitic rocks is a fracture pattern centred on, and concentrated in, apical portions and late phases of the intrusion suite and their immediately superjacent wall rocks: mineralized stockworks, sheeted veins and breccias, i.e. high-permeability zones in a much less deformed regional environment. This focussed permeability pattern, and corresponding centri-symmetric alteration pattern, is typical of copper porphyry systems (Haynes and Titley 1980; Heidrick and Titley 1982), molybdenum porphyries (White et al. 1981; Carten et al. 1988), tin porphyries (Sillitoe et al. 1975; Grant et al. 1980), as well as tin granites (Bolduan 1963; Baumann and Tägl 1963; Teh 1981; Lehmann 1985; Pollard and Taylor 1986). Therefore, the reason for the release of mechanical energy in porphyry-style environments must be related to the emplacement and solidification of certain granitic intrusions, i.e. endogeneous factors, which of course act on the background of a more or less important regional tectonic framework (Barosh 1968; Linnen and Williams-Jones 1987). The endogeneous generation of mechanical energy ($p \cdot \Delta V$) is a natural consequence of crystallization and/or decompression of a hydrous magma in a shallow crustal environment according to the overall second boiling reaction

$$H_2O \text{ saturated melt} \rightarrow \text{crystals} + \text{aqueous fluid}$$

discussed already by Niggli (1920) and quantified more recently by Burnham (1979a, 1985). The increase in volume ΔV in the above equation results from the fact that the partial molar volume of water dissolved in a silicate melt is much smaller than the molar volume of pure water at the same pressure and temperature (for example: H_2O in albite melt = 22 cm^3/mol, pure H_2O = 78 cm^3/mol at 1 kbar and 800 $^\circ$C; Burnham and Davis 1971). In addition, decompression during fracture formation results in further $p \cdot \Delta V$ energy through expansion of already exsolved H_2O in the magma and exsolution of additional H_2O from interstitial melt.

For magmas that initially contained more than approximately 2 wt% H_2O, the decompression and mechanical energy released upon complete crystallization are sufficient to produce intense fracturing in marginal parts of an intrusion and its roofrocks at depths shallower than approximately 5 km

(Burnham 1985). The initial minimum water content of granitic melts is petrologically constrained by the occurrence of biotite and/or hornblende as a phenocryst phase, which both have a lower stability limit of 2-3 wt% H_2O (Wyllie et al. 1976; Burnham 1979b).

The magmatic origin of the early hydrothermal fluids in granite-related ore systems of the Cu-Mo-W-Sn spectrum has been documented by a large number of isotope studies. The hydrothermal evolution of such systems seems to be characterized by progressive dilution by meteoric water. In tin-tungsten deposits, such a trend has been shown by Grant et al. (1980), Patterson et al. (1981), Jackson et al. (1982), Pollard and Taylor (1986), Sun and Eadington (1987) and Thorn (1988) through stable isotope work, and by Higgins et al. (1987) and Higgins and Sun (1988) with the help of Sr and Nd isotopes. The high salinity of the early hydrothermal solutions of Sn-W ore systems as well as the high acid potential of these fluids, which are capable of large-scale pervasive alteration by mainly H_3O^+-consuming reactions, are in accordance with their magmatic origin.

The hydrothermal alteration pattern in tin-tungsten ore systems is essentially identical to copper porphyry systems of Cordilleran type (Hemley and Jones 1964; Rose and Burt 1979; Burnham and Ohmoto 1980). The high concentrations of boron and/or fluorine in Sn-W systems result, however, in the additional and specific alteration components of tourmaline and fluorite ± topaz. A typical alteration sequence is given in Fig. 43. The temporal overprinting sequence comprises potassic, sodic, sericitic and argillic stages in locally very variable proportions, and is paralleled by the mineral sequence of fracture-controlled tin mineralization which in its early phases is associated with feldspar while later phases are dominated by muscovite or chlorite.

In spite of both widespread fluid flow in granitic rocks and the hydrothermal mobility of tin at high temperatures, granitic rocks in association with tin mineralization often have a tin distribution pattern of predominantly magmatic origin, as deduced from the correlation systematics in the examples of the previous chapter. This must be a consequence of incomplete equilibration with a fluid phase (controlled by permeability, flow rate, and kinetics), and is also a result of the relatively lower mobility of Sn when compared with Pb, Zn, and Cu (Crerar and Barnes 1976; Bourcier and Barnes 1987; Wood et al. 1987). The overall hydrothermal solubilities of galena, sphalerite and bornite/chalcopyrite at high temperature in granitic environment are much greater than that for cassiterite, which explains the generally unsystematic and erratic element pattern of base metals in granites.

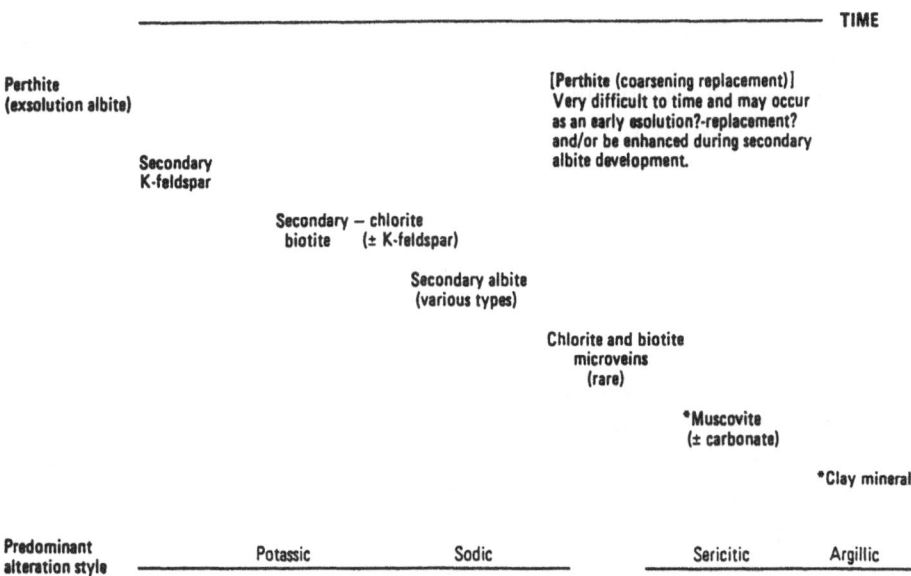

Fig. 43. Timing of main alteration minerals in incipient pervasive alteration of tin granites in the Herberton tin field, Tasmania, Australia. (Pollard and Taylor 1986:1796). This pattern is valid for most tin granites. More advanced degree of alteration results in early formation of tourmaline ± topaz, and blastesis of fluorite over a wide temperature range. Muscovite or zinnwaldite formation begins under early submagmatic conditions

Small-scale metal enrichment in ore deposits and in their hydrothermal haloes must be balanced by zones of large-scale metal depletion. This theoretical condition for ore formation is little documented, and will be addressed in the following examples on tin granites. The most spectacular case of tin depletion is reported for the Tanjungpandan tin granite on Belitung Island (formerly: Billiton), Indonesia. Its anomalously low tin contents were first noted by Cobbing et al. (1986), and subsequent work by Lehmann (1988b) and Lehmann and Harmanto (1990) suggests large-scale hydrothermal tin depletion with an amount of redistributed tin on the order of several million tonnes. The tin depletion zones in the Pilok (Thailand) and Ear Mountain (Alaska, USA) tin fields have much smaller dimensions.

The complementary process of hydrothermal tin enrichment, of which tin ore formation is by definition an integral part, has direct economic impact and is well documented. Our examples are restricted to pervasive tin enrichment patterns and do not include related tin mineralization on major fracture systems. This is, however, an arbitrary distinction which has no point for greisen deposits, in which the process of pervasive alteration and mineralization is identical. Hydrothermal tin enrichment patterns are presented below for tin mining areas in Thailand (Takua Pa), Malaysia (Kinta Valley), and Bolivia (Chacaltaya and Chorolque).

4.1 Tanjungpandan, Indonesia

The Middle Triassic Tanjungpandan batholith on Belitung Island is associated with major alluvial tin deposits and with minor primary tin mineralization of greisen type (Tikus Mine). Subeconomic quartz-tourmaline-cassiterite veinlets and stockworks are locally abundant. The batholith consists of two petrogenetically different, ilmenite-series rock suites with about the same age of 215 ± 3 Ma (Lehmann and Harmanto 1990): an areally dominating biotite granite suite and a quartz syenite suite of more restricted extent (Fig. 44). The granite suite is composed of three subunits which 'are, from oldest to youngest: K-feldspar megacrystic medium- to coarse-grained biotite granite (main phase); megacrystic biotite microgranite with medium- to coarse-grained porphyroclasts of plagioclase, K-feldspar, quartz and biotite (first subintrusion); non-megacrystic biotite microgranite (second subintrusion). The quartz syenite suite covers a large compositional spectrum from gabbroic cumulate rocks to hornblende-biotite quartz syenite (main phase) to alkali-feldspar-hornblende granite pegmatite.

The initial $^{87}Sr/^{86}Sr$ ratio of 0.7140 ± 5 for the granite suite together with an $\epsilon_{Nd}(T)$ value of -7.9 suggest intracrustal source material of Proterozoic age (Darbyshire 1988b; Lehmann and Harmanto 1990). Complete equilibration of major element chemistry at a low-pressure minimum-melt composition, i.e. to the shallow emplacement level (Fig. 45), increasingly negative europium anomalies (Fig. 46), and extended magmatic differentiation trends defined by trace elements (Figs. 47 and 48) indicate the important role of crystal fractionation during the magmatic evolution of the biotite granite suite. Chemical data are compiled in Table 3.

The quartz syenite suite is composed of material with a large spread in Sr initials of 0.7049-0.7153 and corresponding initial ϵ_{Nd} values of -5.7 to -6.1 (Lehmann and Harmanto 1990). This suggests a mixing process between mantle and crustal material. Cumulate textures of the gabbro units and trace element variations in the quartz syenite suite point to magmatic fractionation as an important petrogenetic process.

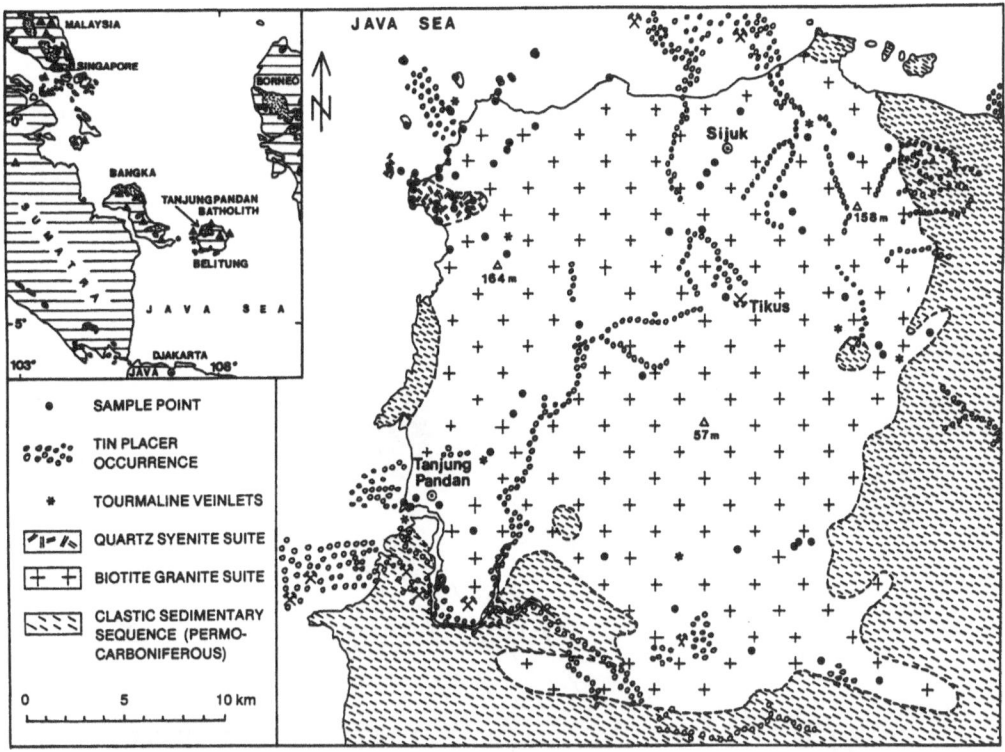

Fig. 44. Location and geology of the Tanjungpandan batholith on Belitung Island, Indonesia. Mine symbols locate past and present tin mining localities. Inset map shows major tin deposits (triangles) in the southernmost part of the SE Asian tin belt, the so-called Tin Islands

The tin content of the biotite granite suite is extremely low, in spite of the high degree of fractionation, the widespread occurrence of quartz-tourmaline-cassiterite veinlets, and the very important alluvial tin concentrations (Fig. 49). The main-phase unit (K-feldspar megacrystic granite) has a mean of 2.8 ppm Sn, the megacrystic microgranite has 4.1 ppm Sn, and the chemically most evolved non-megacrystic microgranite has only 2.1 ppm Sn. The general tin enrichment trend of tin granites, as exemplified by the Erzgebirge granite suite, is included in Fig. 49 for comparison. The average tin contents in the 2-4-ppm range do not satisfy the conventional definition of tin granites which have >15 ppm Sn according to Barsukov (1957).

Contrary to the behaviour of tin, tungsten in the Tanjungpandan rocks is enriched, and gives a scatter distribution with a trend of negative correlation when plotted in terms of the differentiation indicator TiO_2 (Fig. 50), whereas the molybdenum data in Fig. 51 give a trend of decreasing molybdenum content with fractionation. The W and Mo trends are in accordance with the

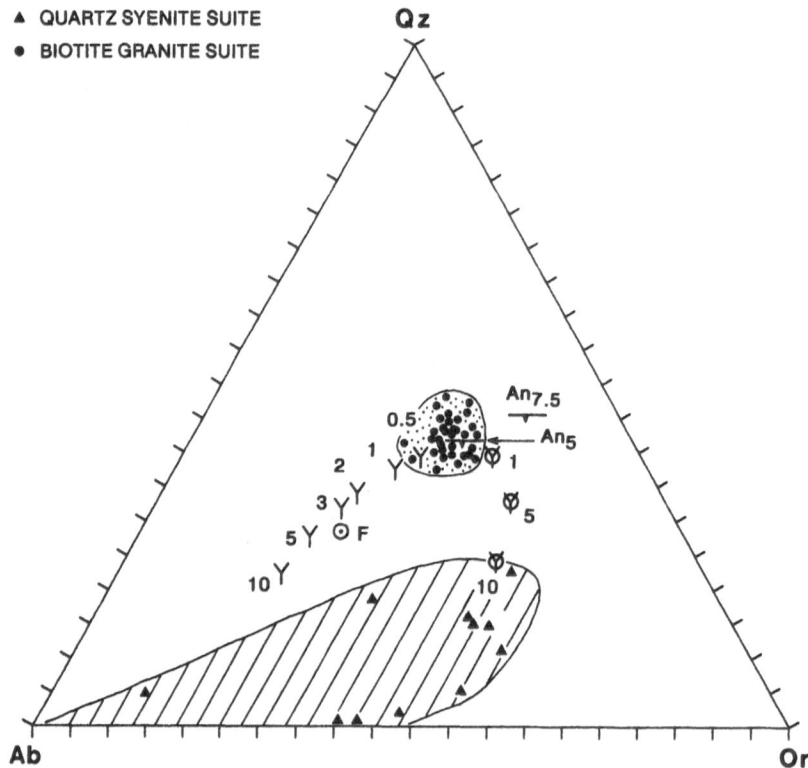

Fig. 45. Normative composition of Tanjungpandan rocks plotted on albite (Ab)-orthoclase (Or)-quartz (Qz) triangle. Y and ❦ symbols locate the experimental hydrous and anhydrous granite minima (Qz-Ab-Or-H_2O system) at various pressures in kbar (Tuttle and Bowen 1958; Luth et al. 1964; Luth 1969), and two minima points of the Qz-Ab-Or-An-H_2O system at $p_{H_2O} = p_{total} = 1$ kbar with An_5 and $An_{7.5}$ (James and Hamilton 1969). Point F indicates haplogranitic minimum-melt composition with 1 wt% fluorine and at 1 kbar (Manning 1981)

experimentally predicted magmatic distribution trends in fractionating low-fO_2 granitic melts, i.e. the dependence of crystal-melt partitioning coefficients of molybdenum and tungsten on oxygen fugacity (Tacker and Candela 1987; Candela and Bouton 1990).

Fig. 46 (next page). REE distribution patterns of the quartz syenite suite and the three subunits of the biotite granite suite, Tanjungpandan batholith, Indonesia

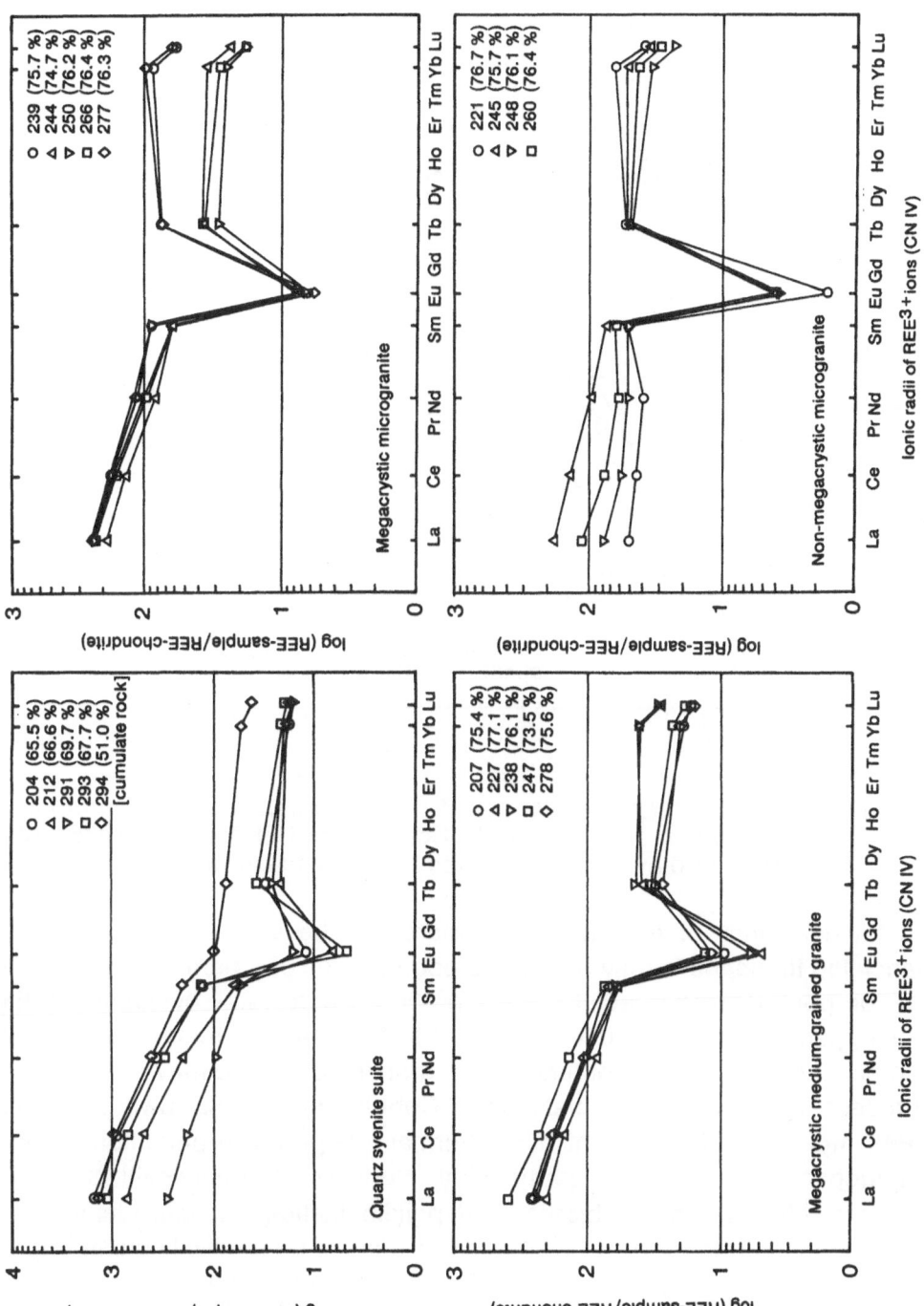

Fig. 46. For legend see previous page

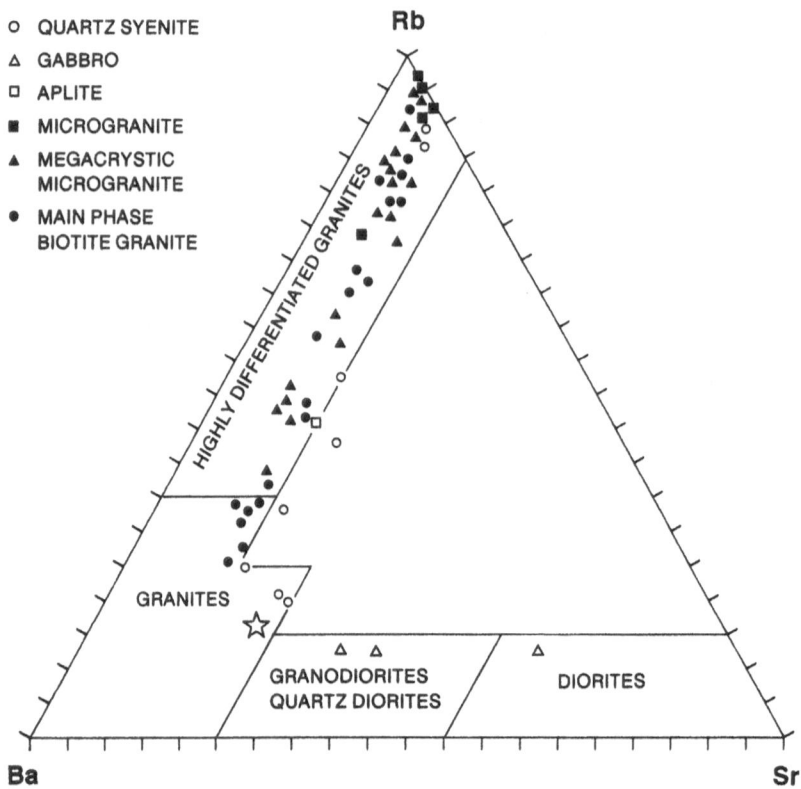

Fig. 47. Rb-Ba-Sr triangle plot of Tanjungpandan rocks. Petrological classification according to El Bouseily and El Sokkary (1975). Star marks world average of granites (Rösler and Lange 1976)

The tin deficiency of the granite compared to the extraordinary wealth of associated tin deposits may be understood in terms of a particularly effective late- or postmagmatic redistribution of tin. The textural evolution of the Tanjungpandan granite suite toward the fine-grained microgranite end-member is a frequent phenomenon in tin granite suites and may be related to a process of quenching by volatile loss (Cobbing et al. 1986; Swanson et al. 1988). Indications of incipient hydrothermal overprint are ubiquitous in the Tanjungpandan batholith; particularly sericitization/muscovitization of feldspars, chloritization of biotite, and poikiloblastic-interstitial blastesis of fluorite and minor tourmaline. Weathering seems to be relatively unimportant, which is also indicated by Fe_2O_3/FeO rock ratios of 0.1-0.01 (Pitfield 1987). These extremely small ratios suggest low fO_2 conditions during crystallization or fluid reequilibration of the granite suite, which is also indicated by the abundance of ilmenite (accessory mineral in the rock and of hydrothermal origin on veinlets) and the occurrence of accessory pyrrhotite.

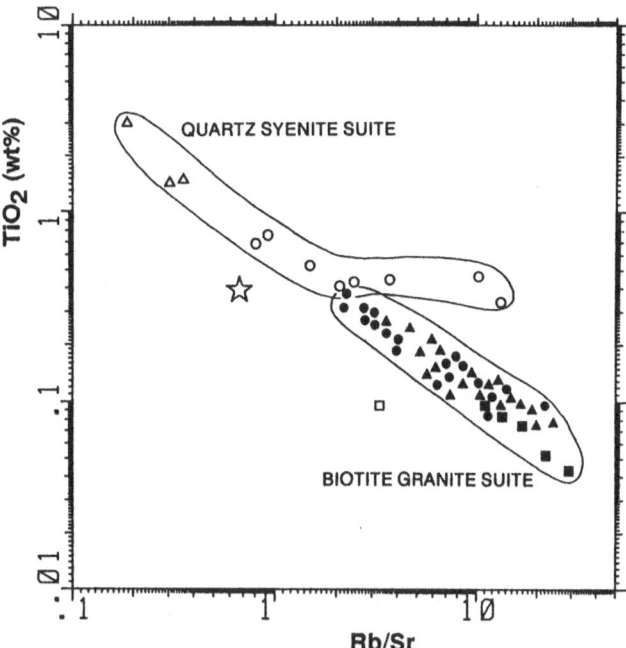

Fig. 48. TiO$_2$-Rb/Sr variation diagram for Tanjungpandan rock samples. Plot symbols as in Fig. 47

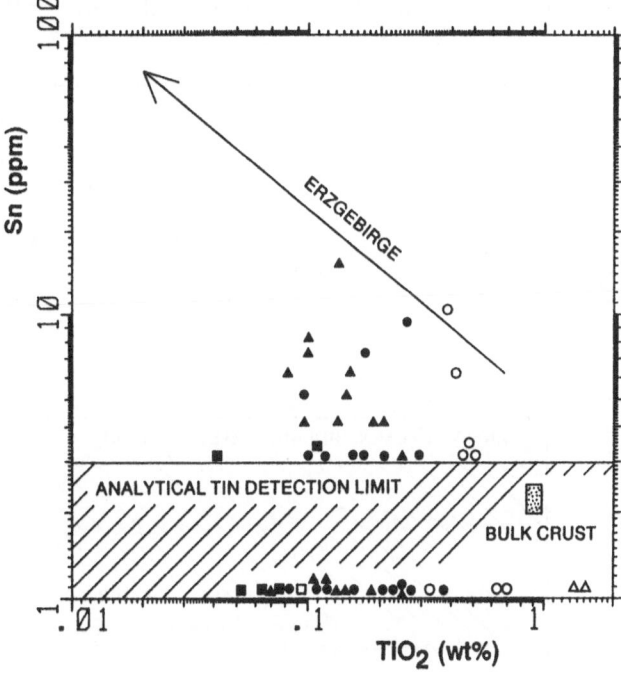

Fig. 49. Tin content in Tanjungpandan rocks as a function of TiO$_2$ (wt%). General magmatic tin enrichment path in tin granites as exemplified by the Erzgebirge trend. Plot symbols as in Fig. 47. Sample points below the analytical detection limit represent values of <3 ppm Sn.

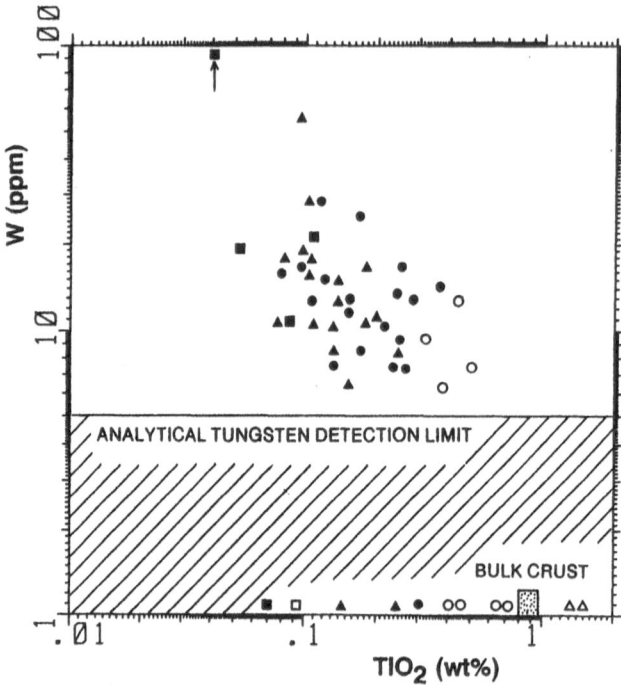

Fig. 50. Variation of tungsten as a function of TiO$_2$ (wt%) for Tanjungpandan rocks. Sample points below analytical detection limit represent values of <5 ppm W. Plot symbols as in Fig. 47.

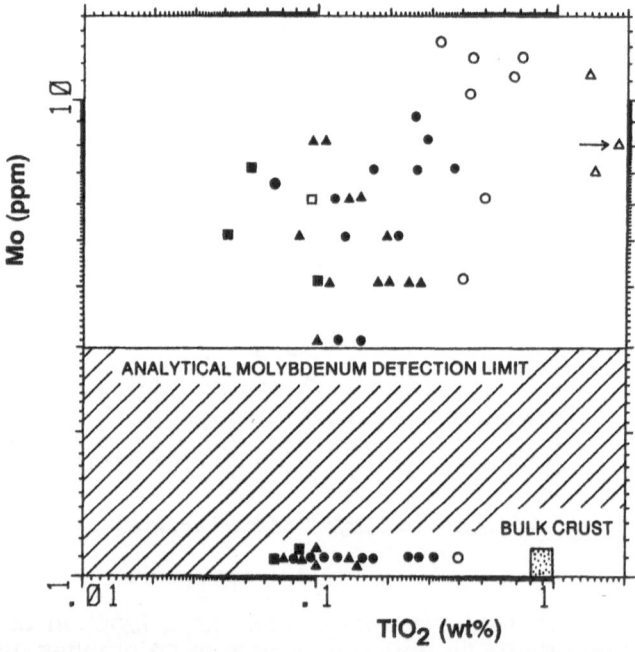

Fig. 51. Variation of molybdenum as a function of TiO$_2$ (wt%) for Tanjungpandan rocks. Sample points below analytical detection limit represent values of <3 ppm Mo. Plot symbols as in Fig. 47

The dramatic dependence of tin mobility on oxygen fugacity has been discussed in Chapter 2.7 and provides an explanation for the exceptionally large-scale tin depletion in the Tanjungpandan granite suite. The neighbouring tin granites on Bangka and Singkep have similar scatter distributions of tin (Fig. 52). These granites have very low Fe_2O_3/FeO rock ratios of <0.1 as well (Pitfield 1987).

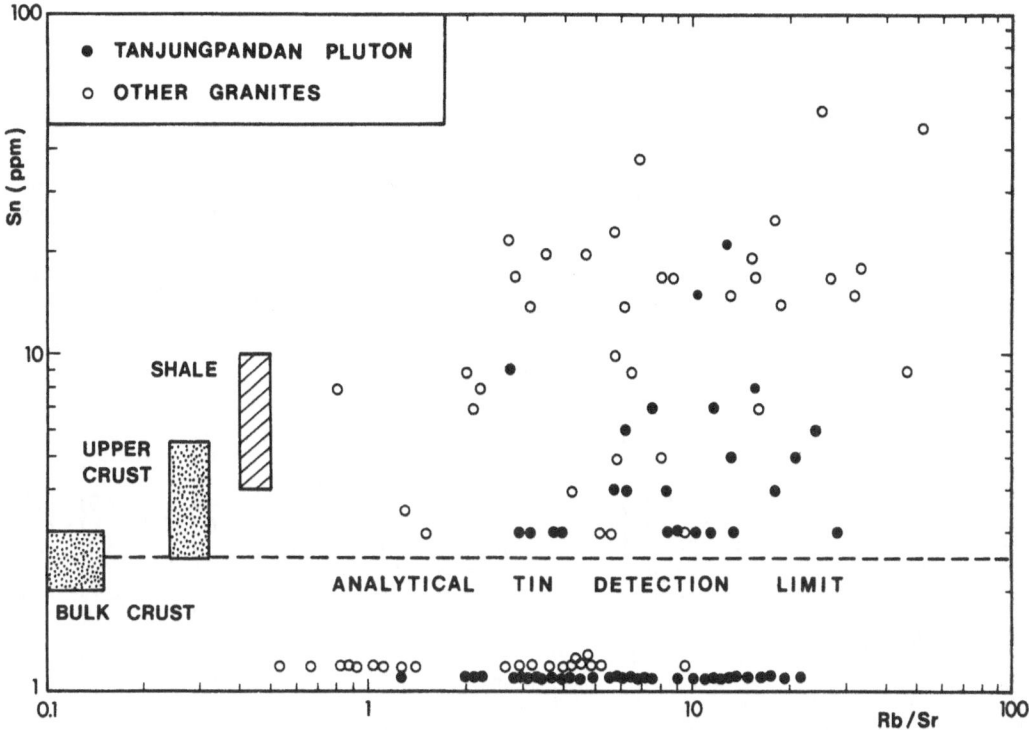

Fig. 52. Tin content as a function of Rb/Sr for Tanjungpandan granitic rocks, Belitung Island, and neighbouring tin granites on Bangka and Singkep Islands. Data from Pitfield (1987)

The primary-magmatic tin content of the Tanjungpandan batholith can be estimated from comparison with the general tin evolution path in less disturbed tin granites (Fig. 49). A conservative estimate arrives at a bulk tin content of around 15 ppm Sn, taking into account the evolved differentiation range of the Tanjungpandan rocks. This figure must be compared with the actual tin content of only 3 ppm. The missing 12 ppm Sn, integrated over the surface exposure of 700 km^2 with an arbitrary depth figure for the hydrothermal depletion system of 500 m, would represent an amount of 10 million tonnes of tin; about ten times the historical tin output of Belitung Island. Of course, only a small part of all redistributed tin is focussed into

98

Table 3. Chemical composition of major rock groups of Tanjungpandan batholith

	Quartz syenite suite		Biotite granite suite			
	Gabbro (n = 3)	Quartz syenite (n = 8)	Main phase granite (n = 19)	Megacrystic microgranite (n = 19)	Micro-granite (n = 5)	Aplite (n = 1)

Major elements in weight percent (mean ± 1 standard deviation):

	Gabbro	Quartz syenite	Main phase granite	Megacrystic microgranite	Microgranite	Aplite
SiO_2	48.65 ± 2.04	65.14 ± 3.02	75.54 ± 1.10	76.10 ± 0.78	76.00 ± 0.67	75.84
TiO_2	1.85 ± 0.86	0.47 ± 0.13	0.18 ± 0.08	0.14 ± 0.05	0.07 ± 0.03	0.09
Al_2O_3	14.72 ± 2.52	16.11 ± 0.98	12.20 ± 0.38	12.06 ± 0.44	12.61 ± 0.25	12.63
Fe_2O_3	10.60 ± 4.31	4.03 ± 1.58	1.79 ± 0.40	1.63 ± 0.29	1.05 ± 0.23	0.71
MnO	0.20 ± 0.06	0.11 ± 0.05	0.04 ± 0.01	0.04 ± 0.01	0.03 ± 0.01	0.02
MgO	6.05 ± 2.80	0.22 ± 0.18	0.13 ± 0.11	0.07 ± 0.08	0.02 ± 0.02	0.04
CaO	11.24 ± 2.72	2.17 ± 0.53	1.29 ± 0.28	1.05 ± 0.15	1.05 ± 0.11	0.75
Na_2O	2.56 ± 0.95	3.62 ± 0.71	2.46 ± 0.23	2.51 ± 0.20	3.01 ± 0.32	2.15
K_2O	1.68 ± 1.29	6.99 ± 0.87	5.23 ± 0.33	5.20 ± 0.29	5.10 ± 0.53	6.72
P_2O_5	0.34 ± 0.13	0.07 ± 0.05	0.03 ± 0.02	0.02 ± 0.01	0.01 ± 0.01	0.02
L.O.I.	1.54 ± 0.72	0.55 ± 0.10	0.65 ± 0.10	0.73 ± 0.12	0.67 ± 0.08	0.74

Trace elements in ppm (mean and range):

	Gabbro	Quartz syenite	Main phase granite	Megacrystic microgranite	Microgranite	Aplite
Ba	403 (94-748)	314 (6-574)	256 (24-568)	134 (<5-418)	24 (<5-104)	265
Ce	107 (73-163)	322 (93-800)	154 (66-228)	122 (84-165)	61 (33-113)	63
Cs	3	3 (1-6)	7 (3-9)	10 (8-12)	6	-
Cu	95 (12-224)	7 (<5-24)	<5 (<5-27)	<5 (<5)	<5 (<5)	<5
F	707 220-1425	273 (110-575)	1662 830-2550	1856 1215-2630	1762 1405-2140	715
La	83 (66-101)	212 (59-532)	95 (54-142)	77 (54-108)	40 (12-68)	25
Li	31	18 (13-22)	35 (22-46)	51 (38-63)	34 (24-44)	-
Mo	8 (5-11)	9 (<3-13)	4 (<3-9)	4 (<3-8)	4 (<3-7)	6
Nb	15 (9-19)	21 (11-31)	15 (12-22)	17 (10-21)	19 (13-30)	6
Ni	40 (10-80)	<5 (<5)	<5 (<5)	<5 (<5)	<5 (<5)	<5
Pb	7 (<5-18)	32 (25-41)	33 (17-48)	40 (26-55)	52 (37-59)	40
Rb	93 (37-155)	228 (173-313)	355 (255-501)	399 (312-504)	408 (356-447)	318
Sn	<3 (<3)	4 (<3-10)	3 (<3-9)	4 (<3-15)	2 (<3-3)	2
Sr	330 (204-490)	124 (16-228)	72 (24-127)	48 (21-104)	26 (15-39)	102
Th	19 (<5-41)	62 (29-107)	75 (52-89)	77 (52-96)	66 (42-108)	25
U	7 (4-10)	10 (6-12)	13 (4-23)	14 (5-21)	19 (15-26)	2
V	212 (93-363)	12 (1-31)	11 (1-26)	9 (4-21)	6 (1-13)	7
W	<5 (<5)	5 (<5-12)	12 (<5-27)	14 (<5-53)	104 (<5-467)	<5
Y	38 (31-44)	43 (32-67)	85 (41-201)	81 (48-167)	114 (79-181)	16
Zn	82 (47-113)	56 (25-91)	24 (16-36)	20 (15-30)	12 (9-18)	16
Zr	219 (161-320)	653 (365-950)	165 (90-251)	145 (101-226)	81 (24-139)	53

	Gabbro	Quartz syenite	Main phase granite	Megacrystic microgranite	Microgranite	Aplite
DI	29.3 ± 13.0	83.4 ± 3.9	89.2 ± 2.1	90.5 ± 1.2	91.8 ± 0.6	93.6
ALUM	0.57 ± 0.14	0.92 ± 0.06	1.01 ± 0.02	1.03 ± 0.02	1.02 ± 0.03	1.04

Note: Besides some complementary F, Cs, Li analyses all data by X-ray fluorescence spectrometry. N is number of samples, Fe_2O_3 is total iron, DI is Thornton-Tuttle differentiation index (normative Qz + Ab + Or), ALUM is index of alumina saturation (molecular $Al_2O_3/CaO + Na_2O + K_2O$).

zones amenable to tin mining. Depending on the local parameters of the hydrothermal circulation system, generally a large part of the redistributed tin is fixed in positive geochemical haloes which have not been sampled in the Tanjungpandan case. Unexposed or little exposed tin granite plutons commonly have a cap of greisen alteration in place, of which the tin-tungsten Tikus deposit is the only known relic in the Tanjungpandan batholith (Jones et al. 1977; Schwartz and Surjono 1988). The erosional level of the Tanjungpandan situation seems to be in an optimal balance between very advanced erosion of lode deposits/primary tin enrichments and a long history of alluvial concentration of cassiterite in residual lag gravel deposits (Aleva 1985).

The exposure of the Tanjungpandan batholith (25 x 30 km in size) at a level perhaps well below the top of the "productive" hydrothermal tin enrichment zone would explain the unusual low-tin granite situation as compared to less eroded tin-bearing systems in Malaysia, Thailand, Burma, Cornwall, Bolivia, etc. which are characterized by high-tin tin granites with average tin contents of generally >15 ppm. The exposure level of the Tanjungpandan granite may therefore provide the rare opportunity to look into the deeper parts of a hydrothermal tin system dominated by tin depletion, with at the same time relics of the upper part of such a system (dominated by tin enrichment) preserved by a fortunate geomorphologic situation of multistage alluvial reworking. The abundant subeconomic quartz-tourmaline-cassiterite veinlets are seen as the root zones of the eroded greisen- and lode-mineralized roof zones.

The reverse case of a particularly effective hydrothermal molybdenum mobilization accompanied by invariable tin levels has been documented by Haffty and Noble (1972) for strongly oxidized rhyolitic rocks in Nevada. Although the complementary behaviour of Mo and Sn as a function of oxygen fugacity was not established at that time, the element patterns in Fig. 53 correspond to the expected trends for hydrothermal alteration at high fO_2 (Fe_2O_3/FeO of the rhyolitic rocks 0.8-2.5; Noble et al. 1968).

Hydrothermal depletion of molybdenum has also been demonstrated more recently for the Miocene rhyolitic ash-flow/dome volcano complex which hosts the Pine Grove porphyry molybdenum deposit in Utah (Keith and Shanks 1988). The ash-flow tuffs are relatively oxidized (log fO_2 reconstructed from Ti-Fe-oxide phases: -16.1 $\pm$ 0.7 at 670 $^{\circ}$C) and show a distinct Mo depletion in even weakly devitrified and hydrated ash samples, with Na content taken as a measure of hydrothermal leaching (Fig. 54). The diagram suggests that molybdenum is leached from the tuff samples in a

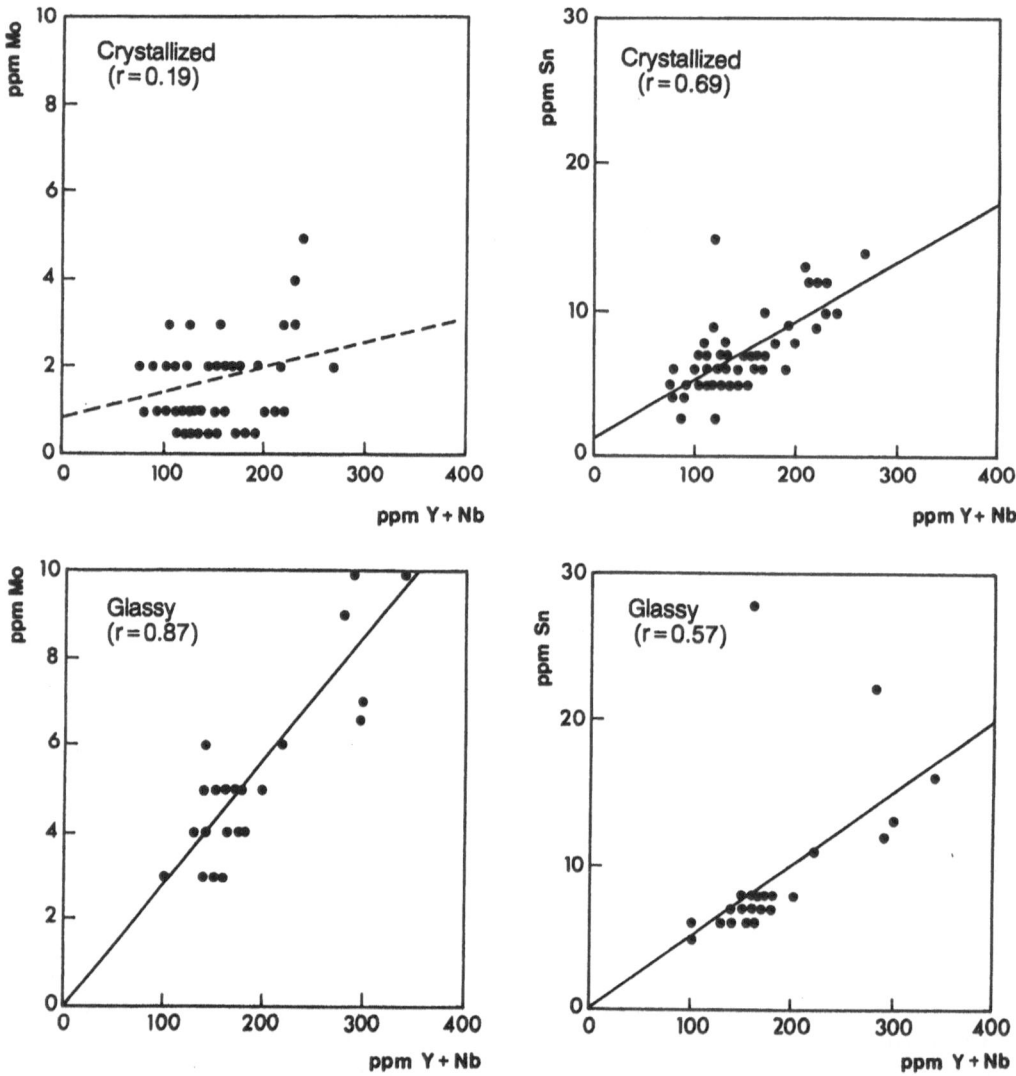

Fig. 53. The different behaviour of tin and molybdenum during devitrification of Tertiary magnetite-series rhyolitic lavas and tuffs in Nevada: constant Sn levels versus hydrothermal Mo depletion. Data from Haffty and Noble (1972). The parameter Y+Nb (ppm) was used by Haffty and Noble (1972) as indicator of fractionation, not affected by hydrothermal overprint. The correlation of Y+Nb vs. Mo and Sn is 0.87 and 0.57 (n=29), respectively, in rhyolitic glass samples, whereas the same correlation in devitrified (crystallized) samples is 0.19 for molybdenum (no significant correlation, i.e. Mo scatter distribution) and 0.69 for tin (undisturbed magmatic correlation trend) (n=57)

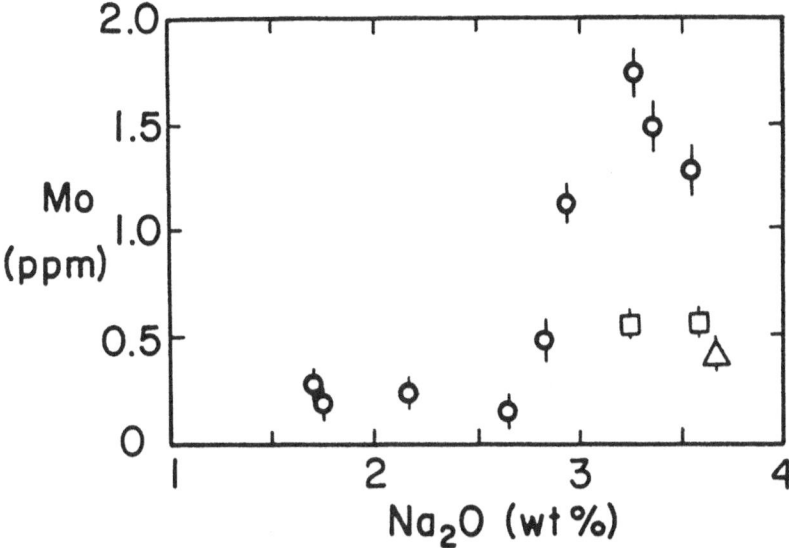

Fig. 54. Whole-rock analyses of molybdenum (ppm) versus Na_2O (wt%) for igneous rocks from the mid-Tertiary Pine Grove volcano-plutonic system in Utah, USA: concomitant leaching of both Na and Mo. Heavy circles represent partly devitrified ash-flow units; the squares and the triangle are rhyolite porphyry samples. Short vertical lines at each data point give analytical uncertainty. (Keith and Shanks 1988:415)

fashion similar to sodium. Samples with magmatic Na concentrations contain about 1.6 ± 0.2 ppm Mo, strongly devitrified samples of the tuff sequence contain only 0.2 ± 0.1 ppm Mo. Fresh porphyry samples contain about 0.5 ± 0.1 ppm Mo, which is thought to be a result of another removal process, i.e. transfer of Mo by an exsolving magmatic fluid phase (Keith and Shanks 1988).

4.2 Pilok, Thailand, and Hermyingyi, Burma

The tin-tungsten mining districts of Pilok in western Thailand and of neighbouring Hermyingyi in Burma (Fig. 31) are centred on highly evolved alkali-feldspar aplogranite stocks of Cretaceous-Tertiary age. The mineralized stocks are exposed in their uppermost portions with surfaces of <1 km^2 and intrude locally foliated K-feldspar megacrystic biotite granite (Border Range main-phase granites) and a Paleozoic sequence of low-grade metamorphic

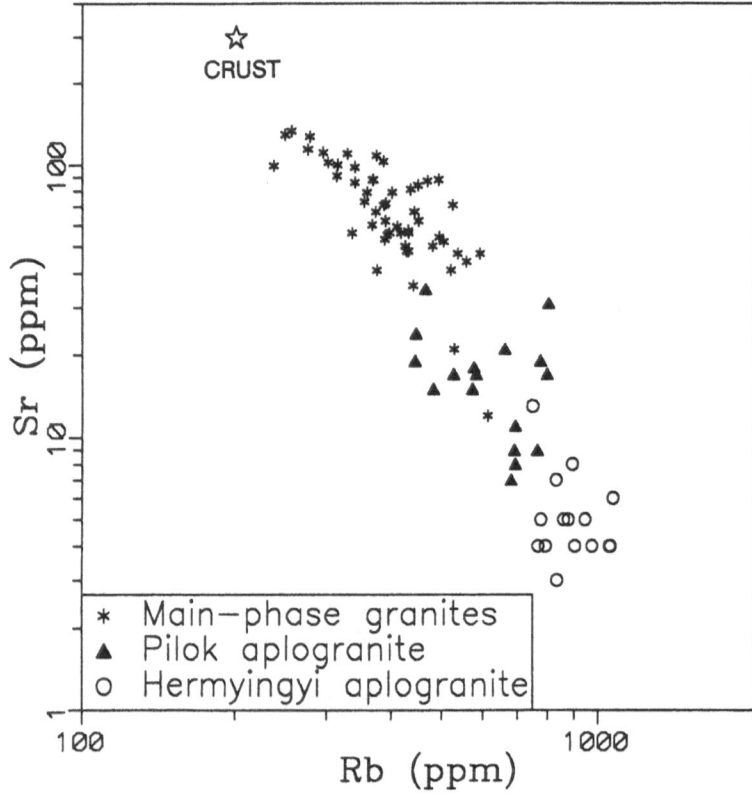

Fig. 55. Rb-Sr variation diagram for granitic rocks from central Thailand and Burma. Open star locates composition of average granitic rocks (Vinogradov 1962, cited in Rösler and Lange 1976). Data from Lehmann (1988a) (Thailand and part of Hermyingyi), and Cobbing et al. (1988) (part of Hermyingyi)

Paleozoic sediments. The alkali-feldspar granite stocks are the product of both advanced fractional crystallization and of fluid overprint in a transitional magmatic-hydrothermal situation; they have the granoblastic (partly symplectic) mineral assemblage quartz-microcline-albite-muscovite-tourmaline-spessartine-fluorite-beryl. The Sr isotope pattern of the Hermyingyi stock is undisturbed and defines a nine-point isochron age of 59 ± 2 Ma and an initial ratio of 0.735 ± 8 (Darbyshire and Swainbank 1988), whereas Sr isotope ratios from the Pilok stock give a scatter distribution and point to exchange with external strontium (Höhndorf unpubl. data). Mineralization consists of disseminations, stockworks, and veins with an ore association composed mainly of arsenopyrite, chalcopyrite, sphalerite, pyrite, cassiterite, and wolframite. Minor components are pyrrhotite, stannite, bismuthinite, bismuth, molybdenite, and scheelite. The gangue assemblage is quartz-

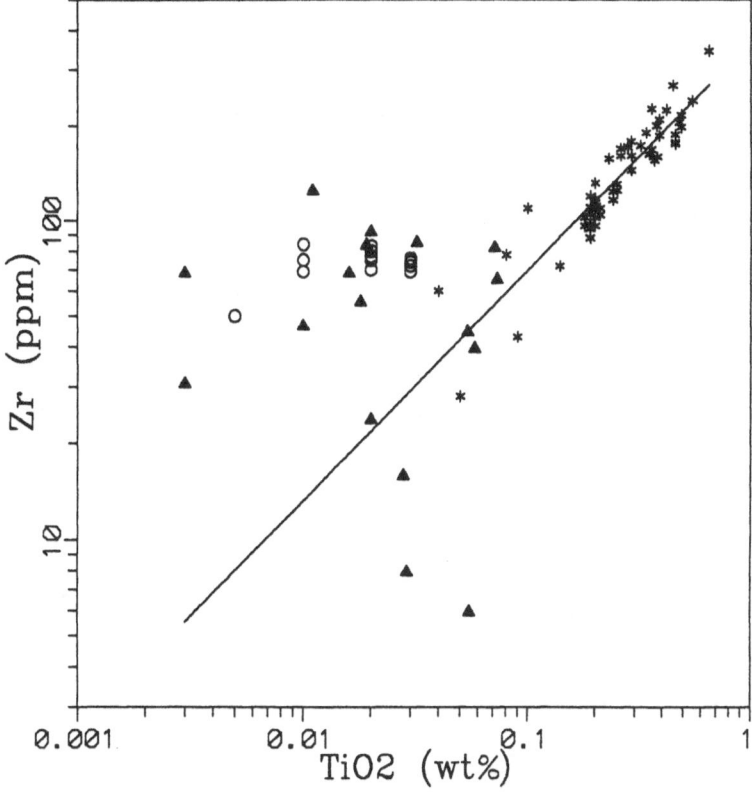

Fig. 56. Zr-TiO$_2$ variation diagram for granitic rocks from central Thailand and Burma. Symbols as in Fig. 55

muscovite-tourmaline-beryl-fluorite-apatite. Pervasive kaolinization is locally strongly developed.

Chemical data of the Pilok and Hermyingyi aplogranites in comparison with granites from the Border Range (main-phase granites of the western granite province), the Rayong Pluton (Main Range province), and the Loei and Chanthaburi granites (eastern granite province) are compiled in Table 4; the general location of these sample groups is given in Fig. 31. Major and trace element concentrations and distribution patterns of the aplogranites indicate an exceptionally high degree of fractionation. The Rb-Sr diagram of Fig. 55 defines a position of the fluid-modified aplogranite samples in the extension of the magmatically controlled trend of the main-phase granites. The Zr-TiO$_2$ variation pattern of the aplogranites is a scatter distribution at the lower part of the extrapolated magmatic trend (Fig. 56). The large scatter implies the absence of any significant control by fractional crystallization of zircon, which in granites is usually an inclusion phase in biotite, or of biotite/Ti-Fe-oxides.

Table 4. Arithmetic means of chemical data for granitic rocks of central Thailand and Burma. Data from Lehmann and Mahawat (1989)

	Loei grano-diorites (n = 7)	Chan-thaburi granites (n = 29)	Rayong granite (n = 25)	Border range granites (n = 20)	Pilok aplo-granite (n = 17)	Hermyingyi aplo-granite (n = 6)
Oxides (wt%)						
SiO_2	63.34	70.28	72.60	73.42	76.18	76.00
TiO_2	0.59	0.41	0.31	0.25	0.03	0.03
Al_2O_3	15.86	14.34	13.80	13.53	13.16	12.83
Fe_2O_3	4.99	3.38	2.08	1.88	0.52	0.99
MnO	0.09	0.06	0.05	0.06	0.07	0.23
MgO	2.20	0.46	0.66	0.35	0.01	0.01
CaO	4.62	2.06	1.26	1.00	0.30	0.50
Na_2O	3.43	3.76	2.74	2.92	3.92	3.45
K_2O	2.99	3.96	5.00	5.11	4.43	4.38
P_2O_5	0.17	0.09	0.16	0.09	0.04	0.01
L.O.I.	1.22	0.68	0.83	0.88	0.73	0.98
Trace elements (ppm)						
Ba	538	368	476	255	13	33
Ce	45	79	66	85	5	47
Cr	29	<15	20	<15	<15	<15
Cu	20	15	6	5	30	18
La	<20	39	26	34	11	81
Nb	7	9	15	24	47	36
Ni	15	8	15	8	9	<5
Pb	11	34	49	87	90	81
Rb	99	207	351	442	625	979
Sn	3	7	10	14	27	72
Sr	381	118	81	65	17	4
Th	10	23	24	43	21	37
U	<5	7	8	19	31	24
V	109	24	25	19	<15	<15
Y	22	51	42	63	102	190
Zn	49	68	32	47	69	135
Zr	147	256	146	149	53	75
D.I.	64	82	86	89	95	93

Note: Analyses by X-ray fluorescence spectrometry. Fe_2O_3 is total iron. D.I. is Thornton-Tuttle differentiation index. L.O.I. is loss on ignition.

Positive correlation coeffcients for Zr with Y (r = 0.60; number of samples = 25), Th (0.73), and Ce (0.55), together with the presence of accessory xenotime and the absence of zircon point to a mineralogical control of the zirconium levels by xenotime. Titanium correlates with Fe (r = 0.61), Mg (0.62), Na (-0.65), D.I. (-0.58) and may therefore be used tentatively as an indicator of fractionation.

The tin content of the rock as a function of TiO_2 and Rb/Sr is shown in Fig. 57. The samples from the Rayong and Border Range granites (main-phase granites) display a systematic trend that can be interpreted in terms of progressive magmatic enrichment of tin which is statistically significant at a confidence level of >99.9% with a correlation coefficient r(logTiO$_2$-logSn) -0.62, and r(logRb/Sr-logSn) 0.65, respectively (50 samples). This is a trend very similar to granite populations from other tin provinces (Lehmann 1982). Three samples of the original sample population (total of 53) are not included in this trend because they have anomalously low tin contents combined with high Rb/Sr and low TiO_2, and plot in the compositional field characterized by fluid interaction. This fluid-controlled field locates all aplogranite samples and is characterized by a scatter distribution in an open system. The tin concentration of these samples is anomalously low relative to their degree of differentiation and is interpreted to be due to removal of tin by fluid interaction. It is further inferred that this mobilized tin is then deposited in fracture systems that are also the main channels for fluid movement.

A geochemical balance of the system can be made when the tin mineralization is taken into account. The original mean content of tin of the Pilok aplogranites can be estimated from the general magmatic correlation trend in Fig. 57. A primary content of 85 ppm Sn in the magma results (standard deviation 35-269 ppm). Compared to the actual mean content of tin, which is 27 ppm (range 0-64 ppm Sn), there is a difference of 58 ppm Sn. This figure provides a basis for estimation of the potential for tin ore of the Pilok

---

Fig. 57 (next page). Tin as a function of TiO_2 and Rb/Sr in granitic rocks from central Thailand and Burma. The main-phase granites (Rayong pluton and Border Range granites) define a tin enrichment trend (solid correlation line) which corresponds to the general magmatic trend for tin granites (Lehmann 1982). Rock compositions in the stippled field (limited by dashed line) are controlled by fluid interaction and are characterized by tin depletion. Note tin deficiency between extrapolated magmatic trend and fluid-modified aplogranite samples. The tin enrichment trends are defined by the correlation lines log[Sn] = -0.86log[TiO$_2$] + 0.57, and log[Sn] = 0.77log[Rb/Sr] + 0.49

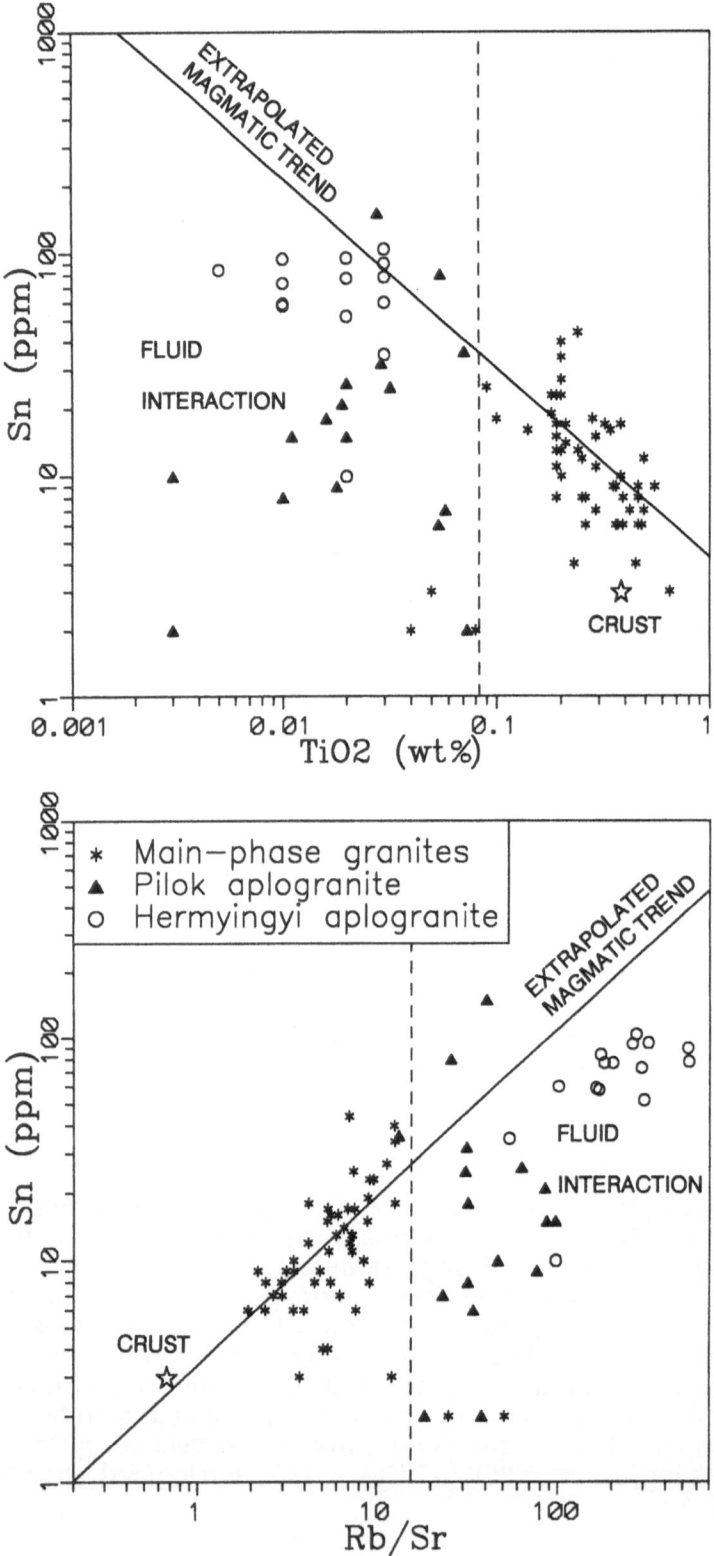

mining area, in which there is a volume of about 1 ± 0.5 km^3 of alkali-feldspar aplogranite. The tin potential of $1.5 \pm 0.7 \times 10^5$ mt (tonnes) Sn, calculated on this theoretical basis, compares reasonably well with the cumulative production plus reserve figures for this area, which total approximately 5×10^4 mt Sn.

4.3 Ear Mountain, Alaska, USA

The 170-km-long, E-W-stretching tin belt on Seward Peninsula in Alaska is defined by nine posttectonic Late Cretaceous granite stocks which are associated with more or less tin mineralization of greisen, vein and skarn type and by placer tin accumulations (Hudson and Arth 1983). The largest primary tin deposit is Lost River with 33 Mio t of ore reserves at 0.29 % Sn (Taylor 1979). The metal potential of this tin province has been estimated at 550,000 t Sn (Reed et al. 1989).

The granites have high intrusion level and are mostly exposed in their uppermost roof portions only. The granite stock of the Lost River Mine has only subsurface outcrops, the Tin Creek pluton has a surface exposure of 0.2 km^2. The intrusions are composed of several texturally different biotite granite units with intrusive contacts. Their age is 70-80 Ma; Sr initials are in the range of 0.708-0.720 (Hudson and Arth, 1983). The granites have peraluminous composition (normative corundum 0.9-2.8 wt%); their composition plots at the low-pressure minimum of the experimental Qz-Ab-Or-H$_2$O system, and An contents are around 2-3 wt%. Chemical distribution patterns have systematics which suggest fractional crystallization as the dominant petrogenetic process during magmatic evolution, with fluid overprint in the most evolved granite phases (Hudson and Arth, 1983). REE patterns of the three major granite variants of the Seward Peninsula emphasize the importance of plagioclase fractionation (Eu/Eu* 0.15 already in the least-evolved granite unit), and of hydrothermal overprint (LREE and Eu depletion) (Fig. 58).

The 9-km^2-large Ear Mountain pluton is located in the middle part of the Seward tin belt and was studied in detail by Swanson et al. (1988). There are four biotite granite phases which define according to field relations and chemistry an intrusion suite of porphyritic → seriate → equigranular → fine-grained texture. The first two granite phases comprise 96 % of the pluton and consist of peraluminous biotite granite. The equigranular and fine-grained variants are biotite-muscovite granite with locally abundant tourmaline and

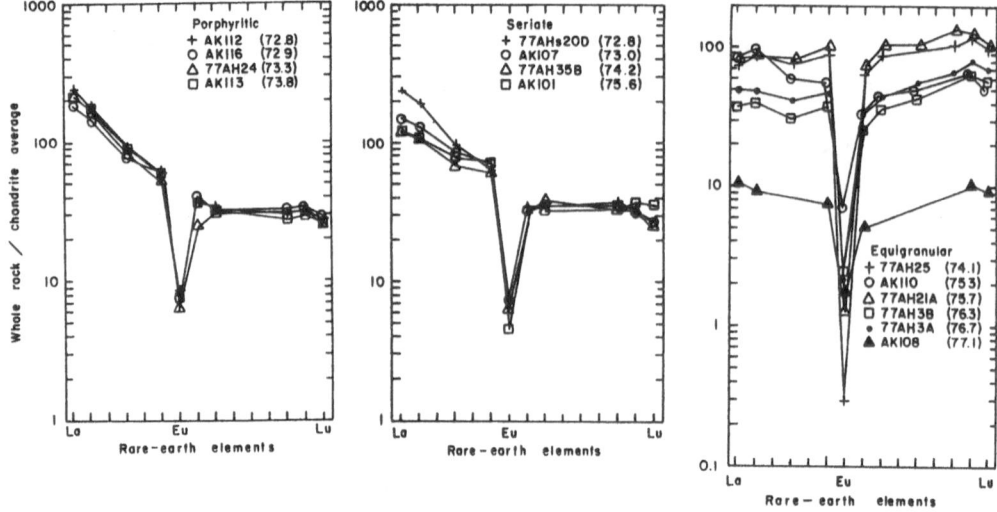

Fig. 58. REE patterns of the three major granite units of Seward Peninsula, Alaska. Intrusive sequence from porphyritic → seriate → equigranular texture. The symbols refer to individual granite samples (SiO$_2$ contents in brackets). (Hudson and Arth 1983)

fluorite, and occur in association with aplite/pegmatite veins and pods. Greisenization is locally and irregularly developed in all granite variants, and is characterized by a recrystallized mineral assemblage of quartz-muscovite-tourmaline with fluorite, cassiterite, rutile, pyrrhotite, pyrite, arsenopyrite and magnetite. Tin contents in greisen zones reach more than 1000 ppm and correlate positively with degree of fluid overprint.

Tin content of the four granite phases is shown in Fig. 59 as a function of TiO$_2$. The two least fractionated granite phases, though much more evolved than average low-Ca granite (Turekian and Wedepohl 1961), follow the general magmatic tin enrichment trend defined by Lehmann (1982). The most fractionated granite phases, on the other hand, are anomalously low in tin. They are modified by high-temperature fluid overprint (blastesis of muscovite, tourmaline, fluorite with feldspar stable), and their tin deficiency can be interpreted as a consequence of vapour exsolution/second boiling, a process which seems to be indicated by the fine-grained quench fabric of these rocks (Swanson et al. 1988). The inferred relationship of such high-temperature tin depletion with small-scale tin enrichment during feldspar-destructive greisenization at lower temperature or with fracture-controlled tin mineralization arises as a requirement of mass balance between the two interrelated systems "granite" and "tin mineralization".

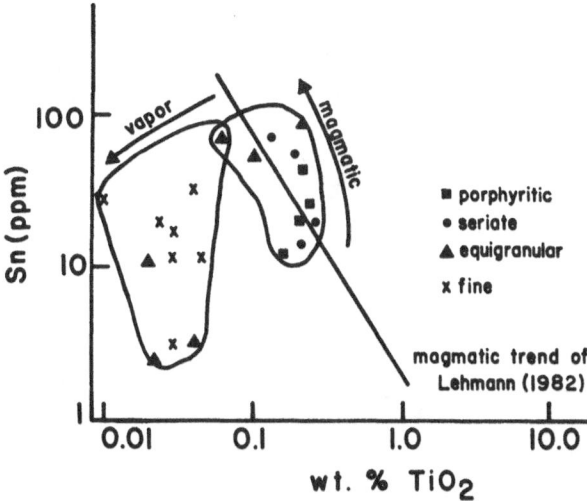

Fig. 59. TiO$_2$-Sn variation diagram for granite samples from the Ear Mountain Pluton, Seward Peninsula, Alaska. Magmatic tin enrichment trend and hydrothermal tin depletion by fluid interaction in the most fractionated granite phases. (Swanson et al. 1988:55)

4.4 Takua Pa, Southern Thailand

The Takua Pa pluton is part of the Phuket-Phangna mining district in southern Thailand which produces the major part of Thailand's tin. The average output of the Phuket-Phangna district over the last twenty years was 4000 t Sn/year, of which most tin is from placer mining. The primary deposits consist of low-grade disseminated greisens, quartz-cassiterite-wolframite veins, and Sn-Ta-bearing pegmatites. Petrography and geochemistry of the Takua Pa pluton was studied by Nakapadungrat et al. (1984a).

The Takua Pa pluton is a posttectonic and discordant, N-S-elongated intrusion (20 x 10 km) in a Permo-Carboniferous weakly metamorphosed and chiefly clastic sedimentary sequence. A four-point Rb-Sr isochron age is 78 ± 2 Ma with $^{87}Sr/^{86}Sr_i$ 0.7346 ± 6 (Nakapadungrat et al. 1984b), identical to the age of the Phuket granites (Putthapiban et al. 1986). The major part of the intrusion consists of several variants of K-feldspar megacrystic biotite granite with variable amounts of sub-solidus muscovite. Younger subintrusions consist of fine-grained biotite granite and of fine- to coarse-grained tourmaline-muscovite granite in which feldspar is partly replaced by muscovite and tourmaline (incipient greisenization).

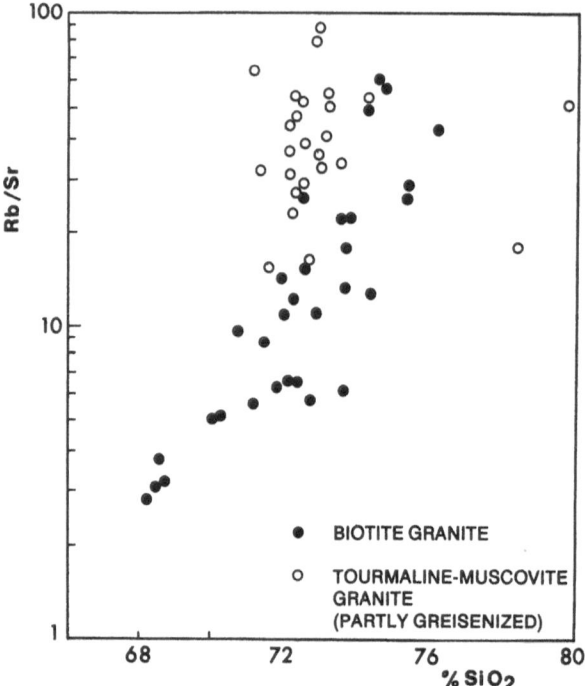

Fig. 60. SiO$_2$-Rb/Sr variation in samples from the Takua Pa pluton, southern Thailand. The biotite granite suite follows a magmatic differentiation trend, the tourmaline-muscovite granite suite has a fluid-modified scatter distribution. Data from Nakapadungrat et al. (1984a)

The normative composition of all rock units is peraluminous (biotite granite suite: 2 ± 1 wt% corundum) and lies close to the thermal minimum in the experimental Qz-Ab-Or-H$_2$O system at 0.5 kbar (An 5 wt%), which indicates advanced equilibration of melt composition with the high intrusion level. Systematic major and trace element distributions point to significant fractional crystallization of plagioclase and K-feldspar (Nakapadungrat et al. 1984a). The Rb/Sr-SiO$_2$ diagram of Fig. 60 gives a positive correlation trend for the biotite granite suite and a scatter distribution for the fluid-modified tourmaline-muscovite granite suite. The tin distribution in Fig. 61 is similarly composed of a systematic tin enrichment trend (dominantly magmatic) and a scatter distribution (fluid overprint) with tin contents up to subeconomic levels in most fractionated granite portions.

Fig. 61 (next page). Tin content as a function of TiO$_2$ and Rb/Sr in the Takua Pa pluton, southern Thailand. Systematic tin enrichment trend in the biotite granite suite with r[logSn-TiO$_2$]-0.58 and r[logSn-logRb/Sr] 0.69 (n=30), and scatter distribution in the fluid-modified tourmaline-muscovite granite suite. Data from Nakapadungrat et al. (1984a)

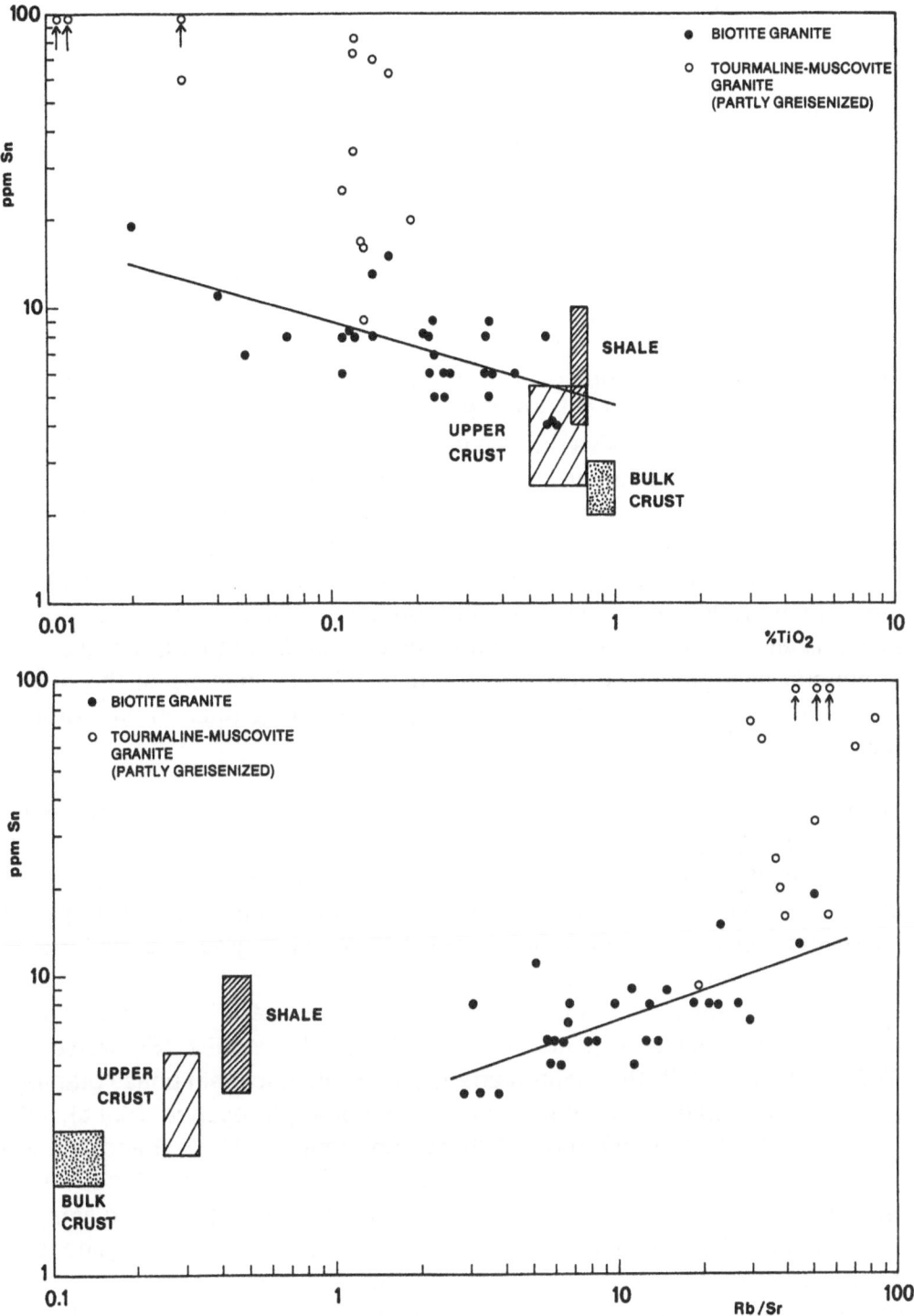

Fig. 61. For legend see previous page

4.5 Kinta Valley, Malaysia

The Kinta Valley with its mining centre of Ipoh is probably the largest tin field in the world and contributes around 30 % to the Malaysian tin production. Tin mining started only in 1876, but production since then has reached a cumulative figure near 2 Mio t Sn, which is about 10 % of the historic world tin output (Rajah 1979). Mining is essentially focussed on alluvial tin, with many small-scale gravel pump operations (up to 500 mines in prosperous times) and several dredges.

The Kinta tin field stretches over an area 90 km long and up to 25 km wide along the broad valley of the Kinta River and is surrounded in a horseshoe-shaped pattern by extensive granitic intrusions of the Main Range batholith of western Malaysia (Ingham and Bradford 1960) (Fig. 62). The local topography is controlled by the contrasting erosional resistivity of a limestone-dominated sedimentary sequence (Silurian to Permian) in the lower parts of the valley, and of Triassic granitic rocks in the surrounding mountain ranges with up to more than 1000 m of relief to the bottom of the valley. Primary tin mineralization of vein, stockwork, and greisen type is frequent; however, it is in general of a grade too low for hard-rock mining. Veinlets of the quartz-tourmaline-cassiterite mineral assemblage are ubiquitous in the granitic rocks.

The topographically striking Bujang Melakka pluton on the eastern edge of the Kinta Valley is an oval (20 x 10 km), N-S-trending granite intrusion with positive relief, surrounded by dozens of gravel pump mines which are mostly located at the deeply weathered granite-limestone contact. The major part of the pluton consists of K-feldspar megacrystic, medium-grained biotite granite with locally variable amounts of sub-solidus muscovite and tourmaline. Chiefly in the central part of the pluton, this granite phase is intruded by non-porphyritic muscovite granite of several textural varieties (Schwartz and Askury 1989). The Rb-Sr isochron age of the porphyritic main phase granite is 207 ± 14 Ma with $^{87}Sr/^{86}Sr_i$ 0.7193 ± 0.0024 (Darbyshire 1988a), fine-grained tourmaline granite has a K-Ar age on biotite of 203 ± 3 and 206 ± 5 Ma (Bignell and Snelling 1977). The Bujang Melakka pluton belongs to the ilmenite granite series (Ishihara et al. 1979; Fletcher et al. 1984) and samples least affected by hydrothermal overprint have a weakly metaluminous to weakly peraluminous composition (Schwartz and Askury 1989). Muscovite granite samples are - by definition - strongly peraluminous, and trace element patterns suggest both a high degree of fractionation and strong hydrothermal overprint.

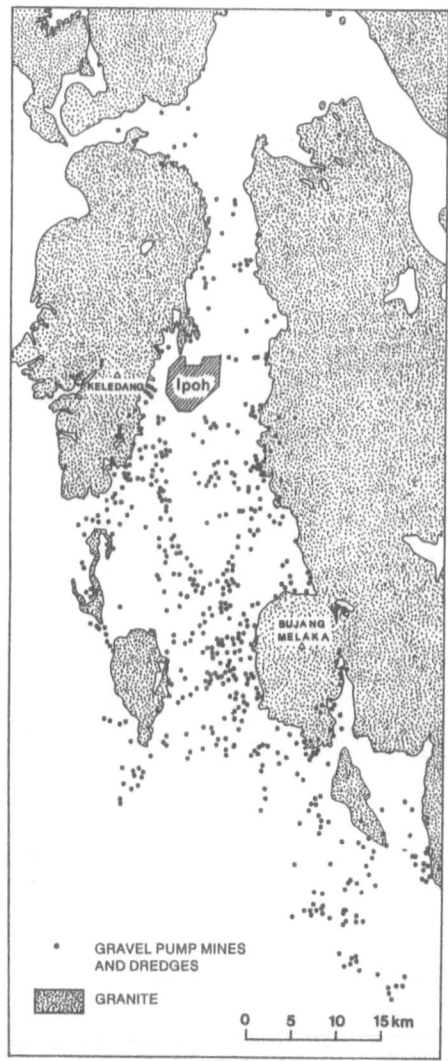

Fig. 62. Active tin mines in the Kinta Valley, Malaysia, during 1982-1985 (Geological Survey of Malaysia unpubl.). The tin deposits consist of alluvial placers and deeply weathered primary deposits, and are mined by hydraulic techniques, mostly small-scale gravel pump mining

The tin distribution pattern of the Bujang Melakka pluton and of some Main Range granites with less hydrothermal alteration is shown in Fig. 63. All granite samples have relatively high tin contents. The biotite granite samples, i.e. samples with low degree of hydrothermal alteration but ± sub-solidus muscovite as minor rock component, give a trend of linear correlation of $\log[TiO_2]$ versus $\log[Sn]$ with a large scatter, which suggests a pattern partly

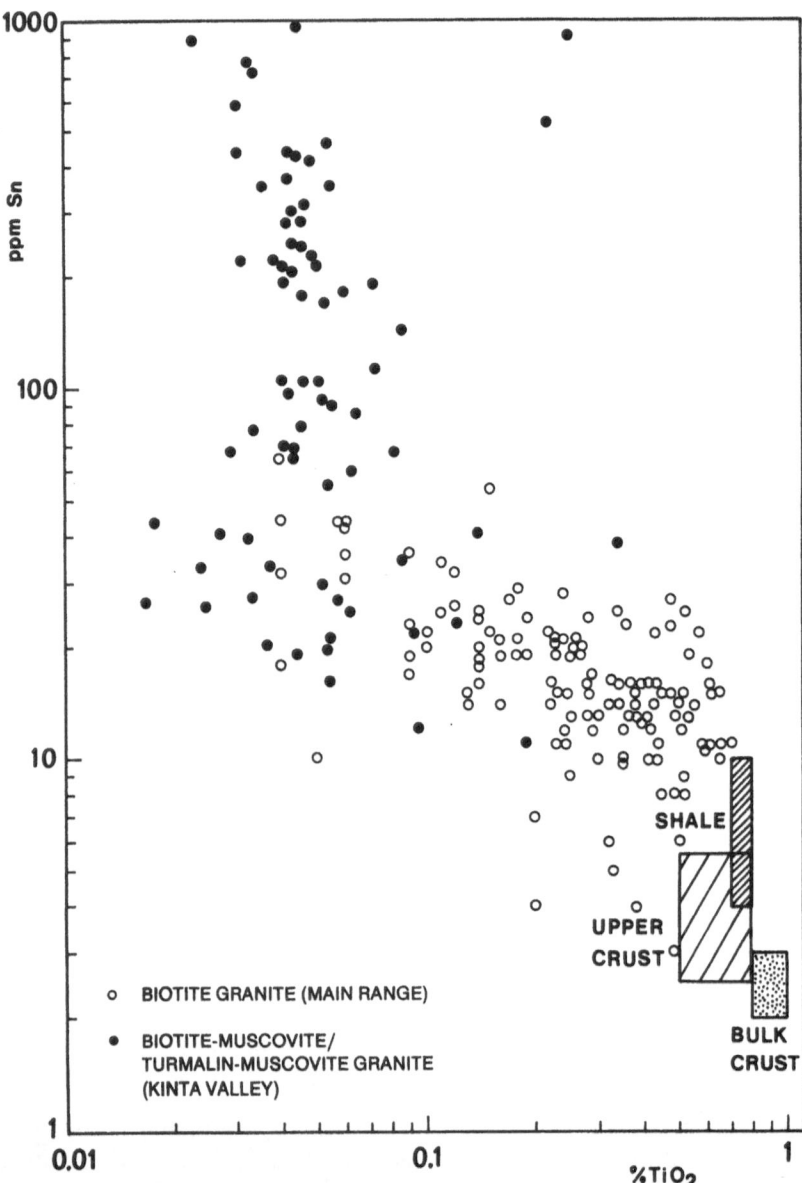

Fig. 63. Tin content as a function of TiO$_2$ (wt%) in granitic rocks of the Bujang Melaka pluton and other Main Range granites in Malaysia. Data from Liew (1983), Schwartz and Askury (1989), Cobbing (1989). Reference rock compositions from Rösler and Lange (1976), and Taylor and McLennan (1985)

controlled by magmatic fractionation and disturbed by secondary processes. A scatter-dominated pattern is developed in those samples which have tourmaline and/or muscovite as a major rock component. These rocks are both highly fractionated (low contents of compatible elements) and are enriched in tin in an open system with presence of a fluid phase. Quartz-topaz greisen bodies reach subeconomic values of 0.3 % Sn and were formerly mined (Ulu Petai Mine; Ingham and Bradford 1960; Schwartz and Askury 1989).

4.6 Chacaltaya, Northern Bolivia

The Chacaltaya granite porphyry with around 0.5 km^2 in outcrop area is located 20 km north of La Paz. It forms the centre of a NW-SE-trending fossil hydrothermal system which is petrographically and geochemically traceable for 12 km in length and includes the mining areas of Milluni and Kellhuani (Lehmann 1985). Tin mineralization occurs in veins and in stratabound stockworks (mantos) which are lithologically controlled by the quartzitic members of a Silurian shale-metasandstone sequence. The Kellhuani mining area has proven plus indicated ore reserves of more than 9 Mio t with 0.5 % Sn (Lehmann 1979).

The Chacaltaya granite porphyry stock has a K-Ar age of 210 ± 6 Ma (McBride et al. 1983), contemporaneous with the Huayna Potosi batholith 8 km further north, and with other tin granites of the northern Bolivian tin belt. The Chacaltaya stock consists of porphyritic biotite granite with sub-solidus muscovite and is extensively affected by hydrothermal overprint. The most advanced product of the hydrothermal alteration spectrum is a heteroblastic, texturally homogenized quartz-muscovite rock without relics of primary-magmatic crystals preserved. These greisen bodies have irregular shape and are several tens of metres in size; tectonic control is not visible. Major rock components are quartz, muscovite, tourmaline, siderite, apatite, fluorite, sphalerite, pyrite, cassiterite and rutile.

The country rocks of the porphyry stock are hydrothermally transformed on a km scale, with a narrow halo of thermal metamorphism preserved relictically in some shale domains (muscovite-biotite-chlorite-quartz-andalusite). The hydrothermal aureole is in its inner part over a length of about 6 km characterized by the granoblastic mineral assemblage quartz-chlorite-tourmaline-fluorite-sericite/muscovite-siderite, the outer aureole differs from the regional-metamorphic reference system by more or less pronounced blastesis of quartz-chlorite-sericite-muscovite-siderite (Fig. 64).

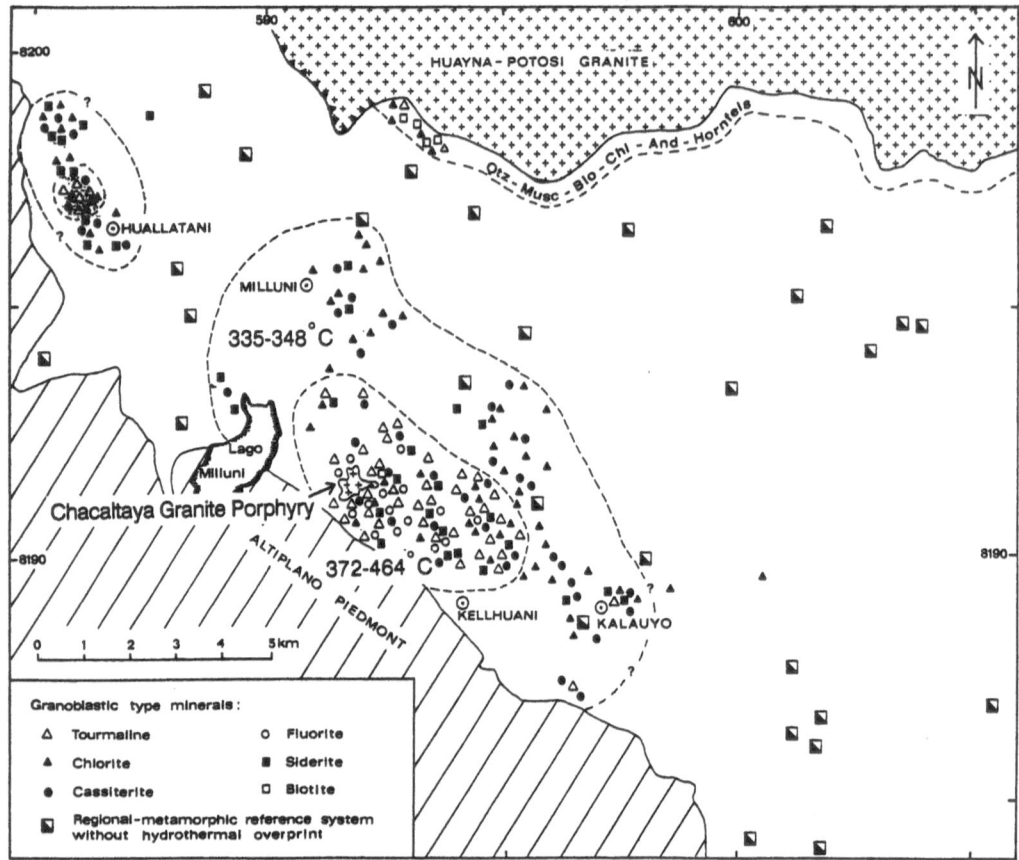

Fig. 64. The hydrothermal-alteration haloes in Silurian quartzites and shales around the Chacaltaya granite porphyry and in the area of Huallatani. The haloes define also the extent of the Kellhuani-Huallatani mining district. Cassiterite homogenization temperatures from Kelly and Turneaure (1970)

The dimensions of the hydrothermal halo are controlled by a fracture pattern of radial configuration which is centred on the porphyry stock and most developed near its contact zone where locally quartz-tourmaline and tourmaline breccias occur. The paleofracture pattern is accentuated by cm-large tourmalinization rims in the wall rocks, and is therefore easily distinguished from younger tectonic elements. The quartzitic fabric is a result of hydrothermal overprint and passes near the halo margin into a sandstone fabric with relatively little blastic grain modification. Stockworks are mostly confined to quartzitic strata, and the manto tin mineralization is located essentially on these closely spaced fractures (Fig. 65).

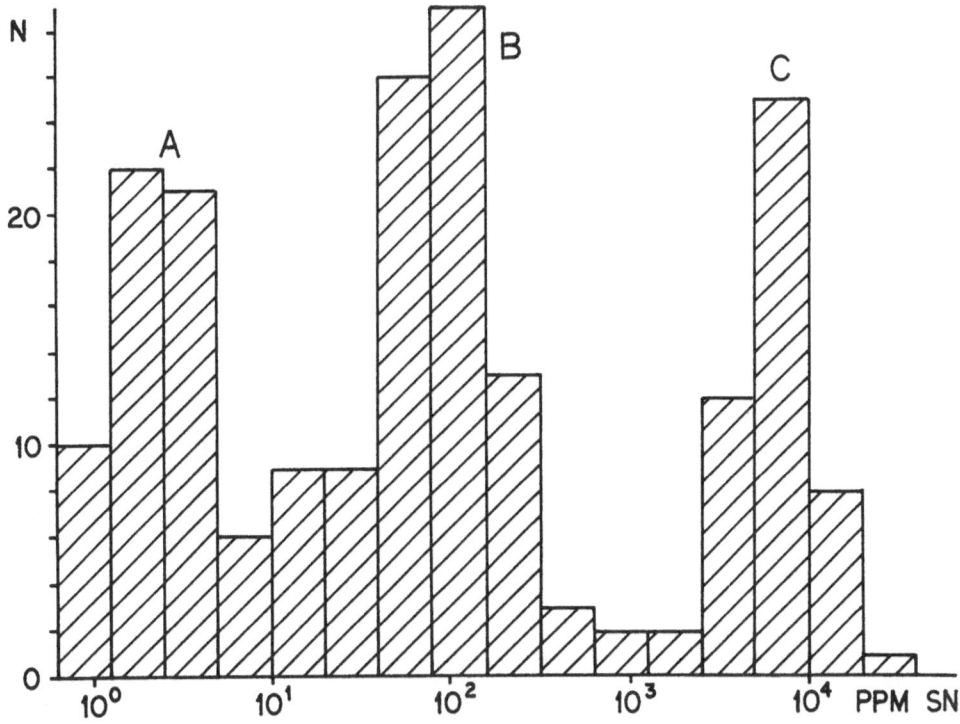

Fig. 65. Trimodal tin distribution in sedimentary rocks of the Kellhuani mining district. **A** regional background (3 ppm Sn); **B** hydrothermal overprint (100 ppm Sn); **C** stockwork/manto mineralization (0.3-1.0 wt% Sn). Data from Lehmann (1979), and Lehmann et al. (1988)

The typical mineral assemblage is quartz-tourmaline-cassiterite. Chlorite, fluorite, siderite, hematite, albite and muscovite are locally abundant. Sulfides are of minor importance; wolframite is restricted to the porphyry and its immediate vicinity. The temperature range of cassiterite crystallization in the Chacaltaya-Kellhuani area is 400-500 °C, in the Milluni vein system temperatures of 380-430 °C are recorded (Kelly and Turneaure 1970; salinity: 5-25 wt% NaCl equivalent; pressure: 0.5-1 kbar; see Lehmann 1985). The polymetallic sulfide mineral association (Zn-Cu-Bi-Pb) of the Milluni veins has a temperature of formation of <380 and >260 °C. In the Kellhuani mantos, the polymetallic mineralization stage is developed on some larger fractures only. Permeability in the stockworks (fracture widths in the mm to cm range) seems

Fig. 66 (next page). Tin contents in granitic rocks from the Chacaltaya and Huayna Potosi intrusions as a function of TiO_2 and Rb/Sr. Data from Lehmann (1979)

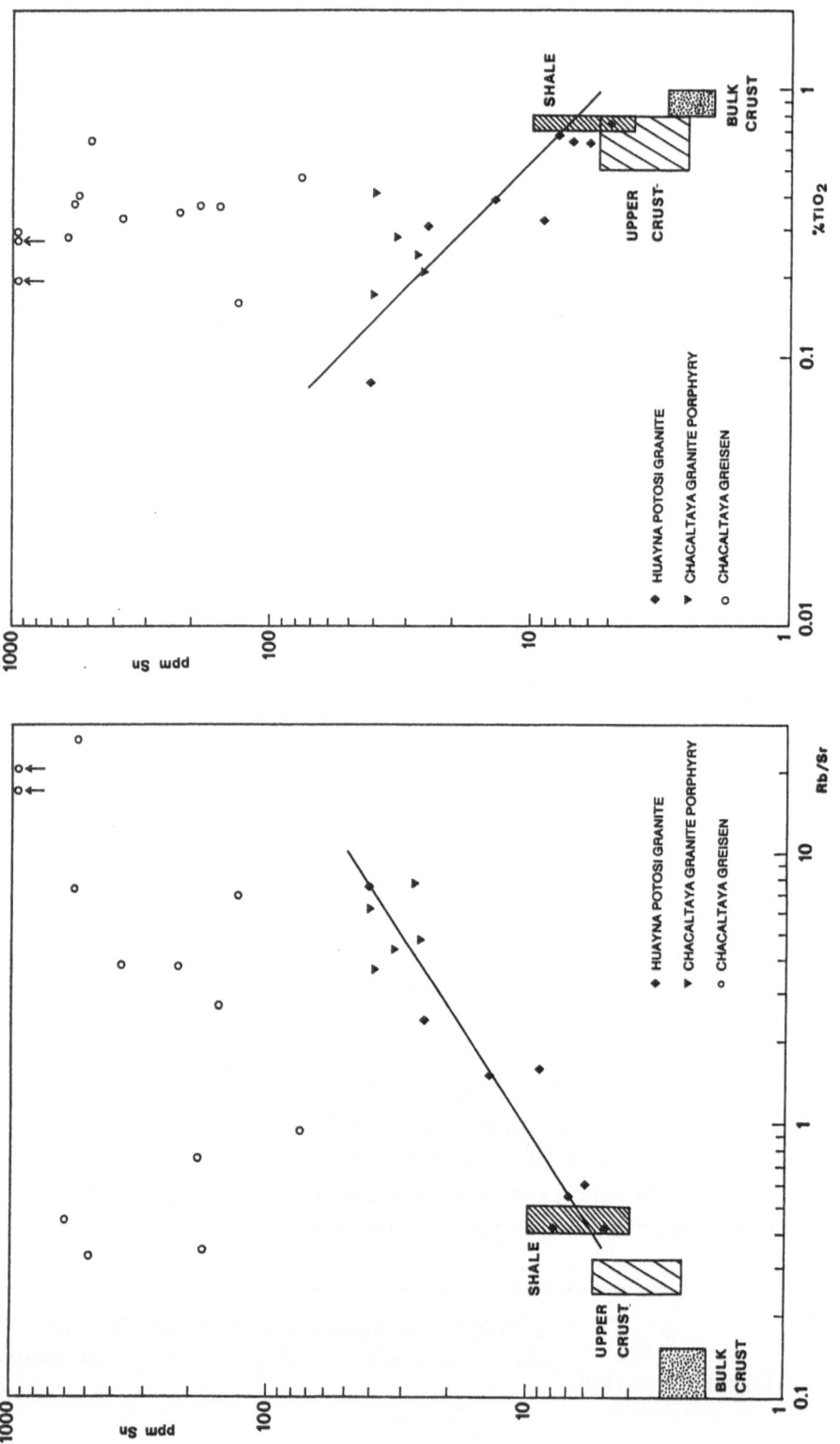

to have been reduced drastically already in the early stages of mineralization, whereas the longer-lived vein structures (dm to m range) allowed mineral deposition over a much larger time and temperature interval. The K-Ar age of hydrothermal muscovite of the manto tin mineralization is 213 ± 5 Ma and is identical to the age of the porphyry intrusion (McBride et al. 1983).

The large-scale halo pattern around the Chacaltaya porphyry can be defined both petrographically and geochemically. Primary dispersion haloes of Sn, F, B, Cs, Li or Zn have the same general configuration but variable, element-specific halo dimensions, with much larger dispersion patterns in the highly permeable quartzite units as compared to the shale units (Lehmann 1979).

There is still a magmatic tin pattern preserved in those samples of the Chacaltaya porphyry which have only moderate hydrothermal overprint. These samples, together with least-altered samples from neighbouring granitic intrusions define a primary tin enrichment trend (Fig. 66) which is

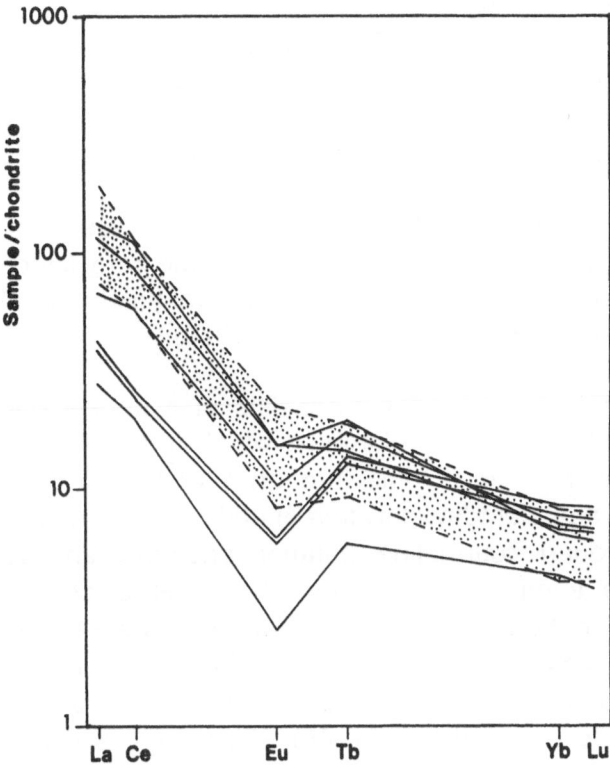

Fig. 67. REE distribution patterns for granitic samples from the Cordillera Real (solid lines) and ten greisen samples from the Chacaltaya granite porphyry (stippled field). Data from P. Dulski, Hahn-Meitner-Institut Berlin

identical to that of the larger Cordillera Real region of northern Bolivia discussed in Chapter 5.3. The strongly hydrothermally altered granite samples plot in a scatter field with generally elevated tin concentrations. Greisenization is not limited to highly fractionated granite portions but affects irregularly the whole range of granite compositions, which may indicate an external fluid source. The greisen samples have a REE distribution pattern identical to equivalent least-altered granite samples from the Cordillera Real (Fig. 67) which suggests very little REE mobility. Eu anomalies are not enhanced by fluid interaction in the Chacaltaya greisens, contrary to the behaviour of Eu during sericitic alteration in the Cornwall tin province (Alderton et al. 1980).

4.7 Chorolque, Southern Bolivia

The Chorolque tin porphyry deposit is located in the southernmost part of the Bolivian tin belt, and has been a tin producer since 1870. Minor amounts of tungsten, bismuth, lead and zinc have been produced from small peripheral veins. Subeconomic silver and gold values are known but are not recovered (Ahlfeld and Schneider-Scherbina 1964). Chorolque is one of the numerous subvolcanic polymetallic mineralization centres in southern Bolivia (Cerro Rico de Potosi, Tasna, Chocaya, Tatasi etc.) which are all of Upper Miocene age with 12-17 Ma, and are thus different from the Early Miocene (>20 Ma) tin porphyry systems further north (Llallagua, Colquechaca, Morococala etc.) which are eroded more deeply (Grant et al. 1979).

The Chorolque igneous complex consists of a largely brecciated circular volcanic vent about 1 km in diameter, which is surrounded by rhyodacitic lavas and tuffs, and which cuts through a peneplain of Ordovician shale basement with a presentday elevation of 4000 m. The vent area is intensively overprinted by quartz-tourmaline alteration, and forms an erosionally resistant peak with the summit at 5600 m. The heteroblastic quartz-tourmaline alteration zone in the central part is surrounded by a sericite-pyrite alteration halo which extends up to 600 m from the intrusive contact into the volcanic and sedimentary rocks and passes gradually into an outer shell of propylitic alteration (biotite → chlorite) (Fig. 68). Tin mineralization is essentially on quartz-tourmaline-cassiterite veins and veinlets confined to the central part of the hydrothermal system. Low-grade disseminated tin mineralization with a polymetallic mineral association of pyrite-arsenopyrite-chalcopyrite-bismuthinite-stannite-cassiterite is located mostly in the peripheral sericite-

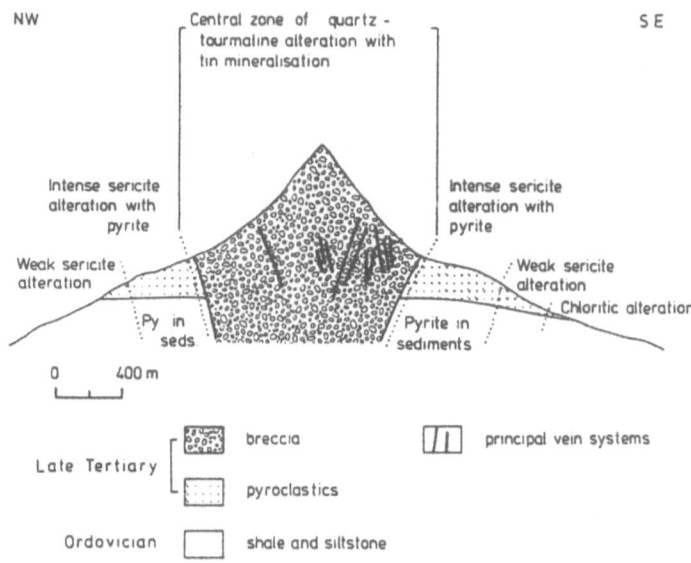

NW

Central zone of quartz -
tourmaline alteration with
tin mineralisation

SE

Intense sericite
alteration with
pyrite

Intense sericite
alteration with
pyrite

Weak sericite
alteration

Weak sericite
alteration

Chloritic alteration

Py in
seds

Pyrite in
sediments

0   400 m

Late Tertiary

breccia

principal vein systems

pyroclastics

Ordovician

shale and siltstone

Fig. 68. Schematic cross-section of the Chorolque tin porphyry and its hydrothermal alteration pattern. (Grant et al. 1977:120)

pyrite alteration zone. The geochemical dispersion patterns of Sn, Ag, Bi and Cu are given in the cross-sections of Fig. 69.

Fluid inclusion studies by Grant et al. (1977) demonstrate that the quartz-tourmaline alteration in the core zone of the hydrothermal system developed at a minimum temperature of 500 $^{\circ}$C and was accompanied by boiling of a saline fluid (co-existence of vapour and liquid phases with >60 wt% NaCl). Cassiterite crystallization is related to lower temperature of 300-250 $^{\circ}$C and lower salinity (<26 wt% NaCl), and is synchroneous with pervasive sericitization. Stable isotope data suggest that this stage of tin mineralization and sericitic alteration was dominated by meteoric water whereas the early high-temperature and high-salinity tourmalinization was dominated by a magmatic fluid phase (Grant et al. 1980).

The temperature-salinity evolution of the tin porphyry system of Chorolque and the much larger Llallagua system are compiled in Fig. 70. Grant et al. (1977, 1980) propose a general genetic model which is very near the model of Burnham (1979a) for copper porphyry deposits: explosive release of an aqueous, Cl- and B-rich fluid phase during the high-level crystallization of a dacitic to rhyolitic melt which leads to pervasive brecciation (hydraulic fracturing) and high-temperature tourmalinization; metal deposition as a consequence of mixing of the magmatic fluid phase with a cooler meteoric

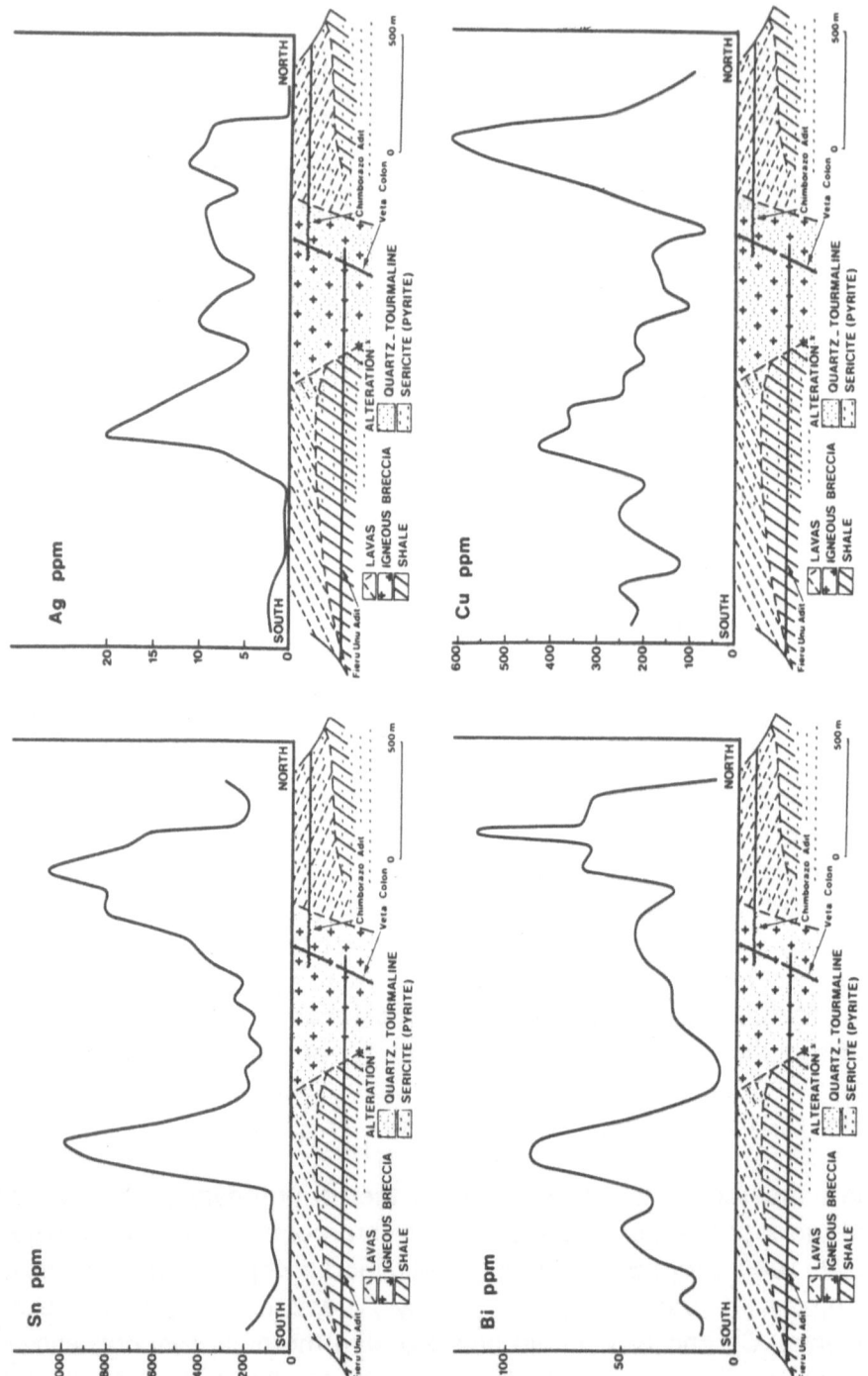

Fig. 69. Geochemical cross-sections through the Chorolque tin porphyry.
(Grant et al. 1977:121)

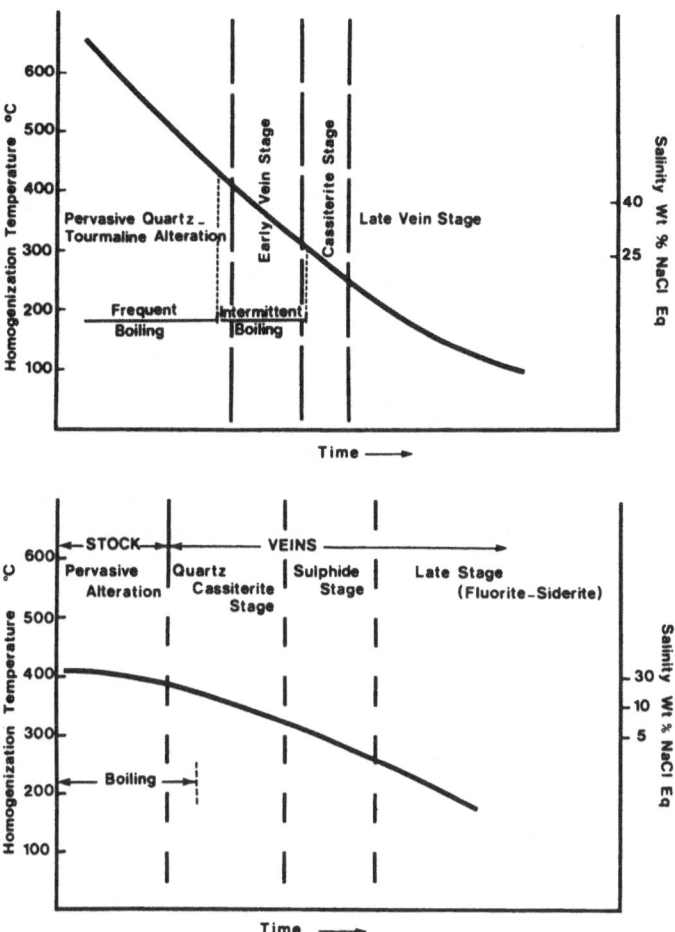

Fig. 70. The temperature-salinity evolution of the hydrothermal systems of the Chorolque (upper diagram) and Llallagua (lower diagram) tin porphyries, reconstructed from fluid inclusion data. (Grant et al. 1977:124)

fluid of low salinity. Recurrent intrusive activity may lead to repetitive patterns of mineralization, with the locus of release of magmatic fluids from a fractionating and crystallizing large-scale magma chamber progressively shifting towards deeper crustal levels.

The magmatic history of the tin porphyries is relatively little known due to pervasive hydrothermal transformations which in many cases do not even permit a reliable petrographic classification of the pre-hydrothermal magmatic material. The magmatic evolution of tin is correspondingly obliterated. This problem will be addressed in Chapter 5.3 on the regional tin distribution pattern of the Bolivian tin province.

Tin porphyries are not limited to Bolivia but are known from the Sierra Madre Occidental in central Mexico, southern China and eastern Siberia as well (Ypma and Simons 1969; Pan 1974; Ruiz 1988; Magak'yan 1968; Guan Xunfan et al. 1984; Yang Shiyi et al. 1984). Subeconomic tin concentrations occur in fluorine-rich, high-silica rhyolites widely distributed over the western USA, known as topaz rhyolites (Burt et al. 1982). The extreme degree of fractionation of the tin-bearing rhyolites of the Mexican tin belt and of the topaz rhyolites in the USA has been documented by Huspeni et al. (1984), Ruiz (1988), and Christiansen et al. (1983). The in general low-grade tin concentrations (≤0.1 wt% Sn) are associated with subvolcanic breccia and stockwork zones in Oligocene rhyolite flows and domes. The locally fayalite-bearing rhyolites ($\rightarrow$low fO_2) are metaluminous to slightly peraluminous, have initial Sr isotope ratios of 0.705-0.708, and are enriched in a suite of incompatible trace elements such as Sn, F, B, U, Th, Rb, Zn, Cs, As (Pan 1974; Burt et al. 1982; Huspeni et al. 1984). The REE spectra in Fig. 71 indicate an advanced degree of plagioclase fractionation.

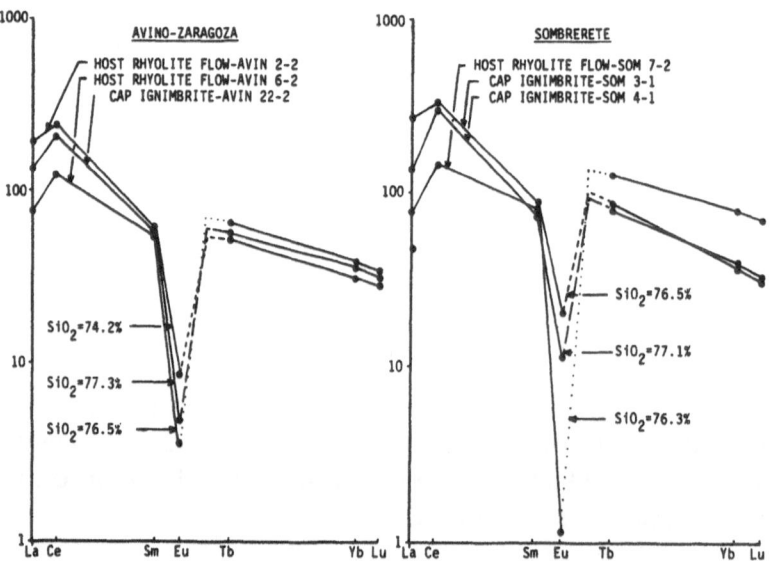

Fig. 71. REE distribution patterns of tin-bearing rhyolites of the Durango tin-silver mining district in Mexico. (Huspeni et al. 1984:102)

5 Regional Element Distribution Patterns and the Problem of Pregranitic Tin Enrichments

5.1 General

The petrological stratification of the lithosphere implies a priori a vertical geochemical stratification on a global scale. Besides this general situation, the existence of metal provinces is often seen from a transformistic point of view as being related to geochemical anomalies of regional extent (Routhier 1980; Schuiling 1967; etc.). Such anomalies should be traceable both by indirect methods such as element evolution trends in granitic differentiation suites as attempted above, and by direct geochemical sampling on a regional scale. Igneous rocks of different origin can be used as probes into their crustal or mantle source regions, high crustal levels can be studied by systematic sampling of outcrops of different lithological units. Such studies need, however, a clear distinction between the regional reference system not affected by hydrothermal alteration and those portions which are hydrothermally overprinted. A metallogenic province of hydrothermal ore deposits contains by definition large-scale hydrothermal haloes, the size of which is sometimes underestimated. Geochemical haloes around hydro-thermal tin ore deposits are commonly several km wide (e.g. Grant et al. 1977; Lehmann 1985; Polya 1988). A mix-up of genetically different sample populations must lead to wrong conclusions.

The literature gives numerous indications of regional metal anomalies (anomalous element contents compared to equivalent rock types elsewhere). Pre-granitic, sedimentary tin and tungsten enrichments have been repeatedly reported from tin and tin-tungsten districts in the eastern USSR and southern China. These data are, however, poorly documented and unverifiable. The problem with such studies is evident from the example of the Erzgebirge.

5.2 Erzgebirge, Germany, and Izera Mountains, Poland

The Proterozoic basement of the Erzgebirge includes a few km northwest of Freiberg a blastomylonitic rock unit with stratabound impregnations of pyrite, known as Felsithorizont. Exploration work in the 1960's established that subeconomic tin contents occur in this rock unit over a strike length of about 10 km. The Felsithorizont of Freiberg-Halsbrücke consists of a heterogeneous sequence of mafic to silica-rich volcanics and of clastic and calcareous sediments, affected by polymetamorphism up to amphibolite grade. Baumann and Weinhold (1963), Baumann (1965) and Weinhold (1977) interpreted the associated tin mineralization as of syngenetic, submarine, hydrothermal-sedimentary formation, which thereby was claimed to be of Proterozoic age with only minor transformation during the Hercynian granite magmatism. This diagnosis of general metallogenetic importance must, however, be rejected in the light of more recent investigations (Kormilicyn 1987; Lorenz and Schirn 1987; Richter 1987).

Systematic studies during the last years demonstrate that the Felsithorizont is a heavily mylonitized sequence of about 1000 m thickness, in which the regional medium-grade metamorphism is reequilibrated to greenschist-facies mineral associations in the most permeable rock portions. Tin mineralization in the form of disseminations is located in metasomatically transformed rock portions; cassiterite-bearing veinlets occur in competent rock units. The general ore assemblage of cassiterite, quartz, pyrite (chalcopyrite and other sulphides are much less abundant), carbonates and fluorite corresponds to the chlorite-sulphide zone of other epigenetic tin deposits of Hercynian age of the Erzgebirge. K-Ar age dating on the retrograde-hydrothermal, chlorite-dominated and mineralized rock portions gave an age of 275 Ma, whereas the non-altered and unmineralized rock portions gave the age of the regional metamorphism of 430 Ma. The unaltered portions of the Felsithorizont have no anomalous tin contents; the contention by Plimer (1980:279) of a "high Sn content of the mafic volcanics of the Erzgebirge" has no basis. A detailed account of this misunderstanding is given by Lorenz and Schirn (1987).

A similar situation seems to apply to strata-bound tin occurrences about 150 km further east in the Karkonosze-Izera Massif at the Czech-Polish border (Fig. 72). There, around the village of Gierczyn, a Proterozoic rock unit about 800 m thick and with a lateral extent of about 35 km is locally associated with tin, copper and cobalt mineralization which was mined in the 16th to 18th century. The ore-bearing horizon (in the German literature known as "zinnerzführendes Fahlband"; Buch 1802; Petrascheck 1933; Putzer 1940, 1942) consists of garnet-mica schist with zones of retrograde chloritization. Mineralization occurs as impregnations and in quartz veinlets in such chloritized zones and consists of cassiterite in association with a polymetallic sulphide assemblage dominated by pyrrhotite, with locally abundant Ni and Co sulphides. The ore-bearing sequence is underlain by the shallow intrusion of the Hercynian Karkonosze granite (around 300 Ma old), which is exposed

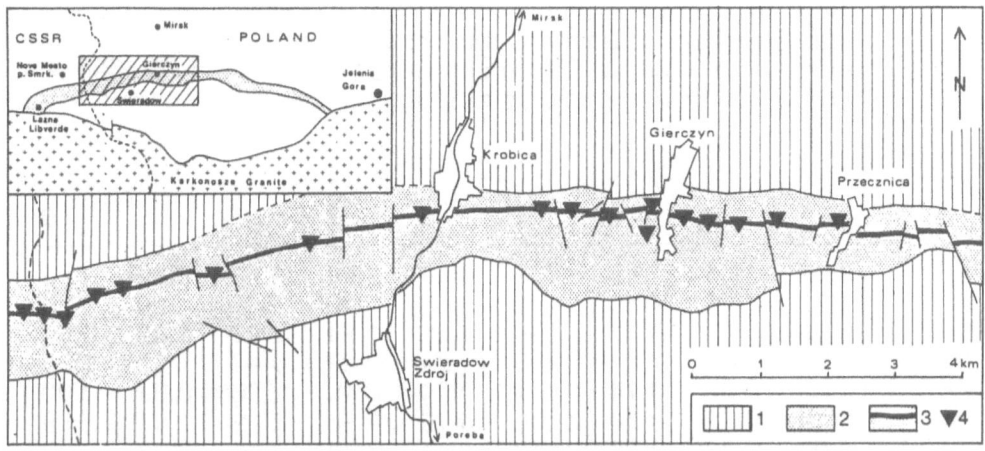

Fig. 72. Generalized geological map of the Gierczyn area, Izera Mountains, Poland. (Lehmann and Schneider 1981:748). **1** augen and flaser gneiss (meta-greywacke); **2** chlorite-mica-quartz schist (+garnet, kyanite, chloritoid); **3** garnet-mica schist (+amphibole); **4** ancient mines (Sn-Cu-Co mineralization)

10 km to the south and which encloses and cuts the ore horizon in a crescent-shaped form (Fig. 72). The Karkonosze granite is host to small tin occurrences in greisen zones (Kozlowski and Karwowski 1975; Kozlowski et al. 1975).

The source material of the garnet-mica schist was a clay-silt-limestone sequence which was transformed by both regional and thermal metamorphism. Volcanic influence is not documented. The conventional concept of a genetic relationship of the stratabound tin mineralization with the neighbouring Karkonosze granite (Berg 1922; Petrascheck 1933; Putzer 1940) was dismissed by Jaskolski (1960, 1962) and Szalamacha and Szalamacha (1974) on the basis of textural-geometric relationships, reviving the conclusions of Buch (1802) from the earliest geological reconnaissance in this area. More recent studies are animated by successful tin exploration in the Gierczyn area and are in favour of a granitic source of the tin mineralization (Kozlowski 1978; Speczik and Wiszniewska 1984). Petrographic and fluid inclusion data define large-scale hydrothermal-metasomatic overprint patterns which have been dated radiometrically with 320-300 Ma (Speczik and Wiszniewska 1984).

Pälchen et al. (1987) calculated regional Clarke values for the pre-granitic basement of the Erzgebirge. These data are summarized in Table 5 and are

Table 5. Regional Clarke   values and regional element abundances in lithostratigraphic units of the pre-Hercynian Erzgebirge (major elements in wt%, trace elements in ppm). Data from Pälchen et al. (1987). Global upper crustal reference values from Taylor and McLennan (1985)

	1	2	3	4	5	6	7	8	9
Si	31.7	29.4	28.6	22.3	40.9	35.0	33.8	30.9	30.8
Al	8.2	10.6	10.9	8.2	3.8	7.0	7.7	9.0	8.0
Fe	3.58	5.07	5.29	8.96	1.83	1.94	1.95	4.22	3.50
Mg	1.09	1.13	1.17	5.6	0.24	0.27	0.43	1.16	1.33
Ca	1.23	0.52	0.14	6.4	0.14	0.46	0.76	0.89	3.00
Na	2.18	0.72	0.70	2.00	0.15	1.93	2.19	1.42	2.89
K	3.05	3.31	3.26	0.42	1.41	3.74	3.40	2.99	2.80
B	13	34	65	<10	46	12	15	32	15
Ba	660	515	675	80	355	160	490	565	550
Be	2.9	2.3	3.0	1.0	1.0	2.0	1.9	2.4	3.0
Co	11	11	14	49	5.3	2.0	2.7	11	10
Cr	48	65	72	240	21	24	18	60	35
Cu	25	26	26	60	14	9.0	9.3	26	25
F	610	625	765	450	210	600	685	615	720
Li	66	53	84	35	28	30	46	62	20
Mn	460	485	1050	1450	155	160	205	575	600
Ni	20	23	39	72	13	4.0	6.0	26	20
Pb	24	23	21	8	16	30	25	22	20
Rb	115	120	200	30	64	230	220	140	112
Sn	3.9	3.7	5.0	4.0	2.0	8.0	5.9	4.6	5.5
Sr	175	85	89	200	34	35	68	115	350
Ti	4150	4750	5600	11500	4050	1000	1500	4500	3000
V	68	91	115	300	29	12	18	80	60
W	1.8	1.6	1.9	1.0	1.0	3.5	4.2	2.4	2.0
Zn	80	77	115	90	30	40	55	83	71
Zr	205	240	160	110	730	60	105	195	190

1 = Paragneiss (n=32)
2 = Mica schist (n=25)
3 = Phyllite (n=353)
4 = Meta-basite (n=22)
5 = Quartzite (n=91)
6 = Meta-rhyolite ("Graugneis") (n=55)
7 = Meta-granitoid (n=197)
8 = Regional Clarke of pre-Hercynian Erzgebirge
9 = Clarke according to Taylor and McLennan (1985)

derived from a large number of representative rock samples which were carefully examined for absence of any signs of hydrothermal overprint. The average element contents of tin and tungsten, as well as of most other elements are very close to global averages for the upper crust. There is no indication for a pre-Hercynian tin enrichment such as a regional geochemical tin anomaly in the Erzgebirge basement.

This result contrasts with an earlier attempt by Weinhold (1977) to define regional element abundances in the metamorphic basement of the Erzgebirge. This study arrived at an arithmetic mean of 68 ppm Sn, based on 1725 rock samples. Numerous hydrothermally altered samples were, however, included in this calculation, which could have easily been identified by analysis of the frequency distribution of the data. The data evaluation by Weinhold (1977) is considered today as erroneous (see discussion in Pälchen et al. 1987) and illustrates the fact that any investigation on regional element abundances needs careful sample selection and verifiable data processing.

5.3 The Bolivian Tin Belt

The enormous boron and tin accumulations in the hydrothermal systems of the Bolivian tin belt could be derived from sedimentary tin and boron anomalies in the thick Lower Paleozoic sequences of that region, remobilized by hydrothermal convection cells associated with the Triassic and Tertiary magmatism (Petersen 1979). The country rock of both Triassic tin granites and Tertiary tin prophyries consists of a Lower Ordovician to Upper Devonian clastic sequence (shale, meta-sandstone and siltstone), with more than 10,000 m of stratigraphic thickness. This sequence is the result of intracratonic, marine sedimentation with source regions dominantly to the west of the shallow basin in the "Altiplano Massif" (Isaacson 1975), and has been folded and metamorphosed under very-low- to low-grade conditions during the Hercynian orogeny (Martinez 1980).

The Lower Paleozoic sequence has been sampled in a cross-section between La Paz and Coroico, about 50 km long and perpendicular to the strike of the eastern Andes (Lehmann et al. 1988). The 120 samples (1-2 kg each) from Middle Ordovician to Upper Silurian shales and meta-sandstones give mean values of tin and boron contents which correspond to averages of equivalent rock types from other parts of the world (Fig. 73). These data do not support the assumption of a regional tin or boron anomaly in the Lower Paleozoic

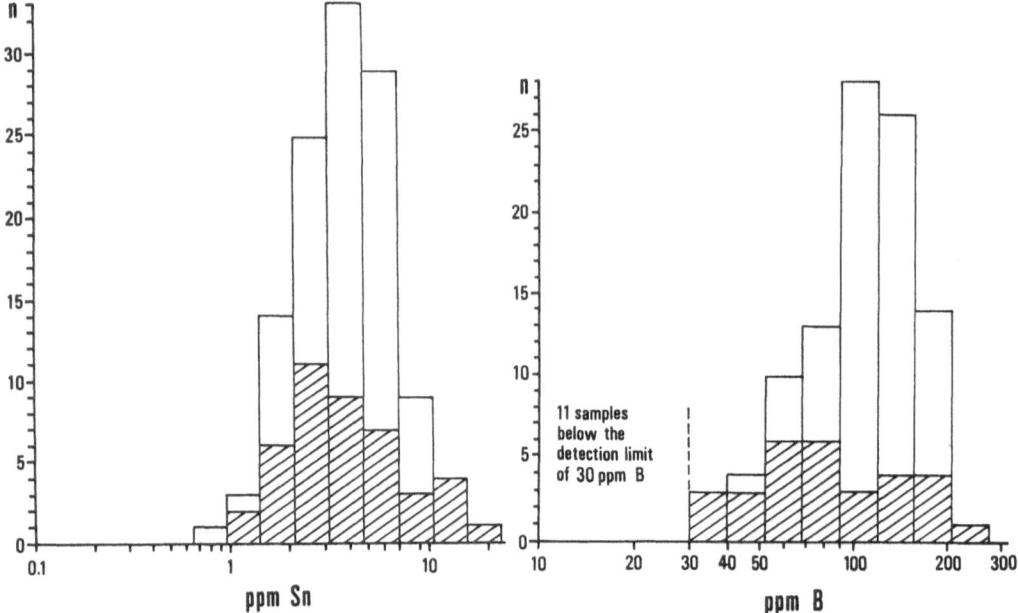

Fig. 73. Regional tin and boron abundances in the Ordovician-Silurian shale-sandstone sequence of the Cordillera Real, Bolivia. Sample section La Paz-Coroico; 110 rock samples. (Lehmann et al. 1988). Ruled: meta-sandstone; open boxes: shale. Eleven samples with <30 ppm B are meta-sandstones

sequence of the Bolivian tin belt. However, there is still the possibility that hypothetical pre-Hercynian tin and boron accumulations may be unexposed or may have gone undetected.

Lead isotope data on galena from some Bolivian tin deposits indicate an origin predominantly from crustal sources (Tilton et al. 1981). A discrimination between Paleozoic wall rocks or granitic intrusions and their Precambrian source material is, however, not possible.

Although the available geochemical data make a geochemical tin anomaly in the Paleozoic wall rocks of the tin-bearing igneous intrusions unlikely, there is, of course, the possibility that the Triassic and Tertiary intrusions are derived from partial melting of deep and unexposed Precambrian tin anomalies (Schuiling 1967), or that even a subcrustal tin anomaly is involved (Clark et al. 1976). A first regional study on the tin distribution in mid-Andean volcanic rocks by Lehmann and Pichler (1980) reported mean values of 2 ppm Sn in rhyolitic rocks of northern Chile and NW Argentina, 3 ppm Sn in rhyolitic rocks from the Bolivian Altiplano, and 4.5 ppm Sn in rhyolites and rhyodacites of the eastern Cordillera of central Bolivia (Los Frailes Formation). These data

Fig. 74. Location of sample groups discussed in text (stippled areas), and of major copper porphyry and tin ore deposits in the Central Andes

appear to define a positive regional tin gradient towards the Bolivian tin belt. This picture can be further accentuated by new data from Winkelmann (1983), Gardeweg et al. (1984), Ishihara et al. (1984), Tistl (1985), Lehmann (unpubl.) and Miller (1988).

Tin data on igneous rocks from the following areas in the greater region around the Bolivian tin belt are available (Fig. 74):

1. Granitic rocks from northern and central Chile (Antofagasta and Copiapó sections), subdivided into a Paleozoic and a Mesozoic-Cenozoic group; sample locations and petrographic and chemical data are given in Ishihara et al (1984). The samples of the Paleozoic group are dominantly biotite granites with initial Sr isotope ratios ≤0.706 (Shibata et al., 1984), and having a petrochemistry of partly S-type and partly I-type. The Mesozoic-Cenozoic group consists of a calcalkaline suite of predominantly quartz-monzodioritic to granodioritic composition and is associated with the famous Chilean copper porphyries. These granitoids are hornblende-and magnetite-bearing and have initial Sr isotope ratios around 0.704 (I-type; magnetite series) (Gustafson and Hunt 1975; Shibata et al. 1984).

2. The granitic rocks of the northern Bolivian tin belt are subalkaline and peraluminous biotite granodiorites (hornblende-bearing) to biotite-muscovite syenogranites. K-Ar age data define two granite populations: a Middle to Late Triassic group (225-202 Ma) and a Late Oligocene to Early Miocene group (28-19 Ma) (McBride et al. 1983). Both granite groups are locally associated with tin-tungsten deposits, and have been sampled - from north to south - in the Huato, Illampu, Zongo-Yani, Huayna Potosi, Chacaltaya, Unduavi, Taquesi and Chojlla intrusions, as well as in the Quimsa Cruz batholith (Lehmann 1979; Winkelmann 1983; Tistl 1985; Miller 1988). The Triassic granites in northern Bolivia are part of a 1200-km-long Permo-Triassic granite belt which extends to central Peru, and are interpreted as rift-associated intracratonic magmatism (Kontak et al. 1984). Initial Sr isotope data from southern Peru are in the range of 0.7081-0.7170 (Kontak et al. 1984), preliminary data from the Cordillera Real give 0.7079-0.7087 (McNutt and Clark 1983). The Zongo-Yani pluton is locally foliated and has a synkinematic metamorphic aureole attaining sillimanite grade (Bard et al. 1974). Recent Rb-Sr isotope data indicate a Permo-Carboniferous age (Harris 1988), and $\epsilon_{Nd}(T=284$ Ma) -6.0 (Miller and Harris 1989). The Tertiary granites in northern Bolivia (Quimsa Cruz intrusions) are the more deeply eroded equivalents of the tin porphyry systems of central and southern Bolivia, with which they form an extended volcanoplutonic province related to the subduction of the Nazca plate under the South-American craton.

3. The Ordovician Cafayate and Cuchiyaco granites of the Pampean Ranges in northwestern Argentina are peraluminous, calcalkaline fractionation suites which range from biotite tonalite to biotite-muscovite granite (Rapela and

Shaw 1979; Rapela et al. 1982). These rocks have an initial Sr isotope ratio of 0.705; they are unmineralized (Saavedra et al. 1987).

4. The Cenozoic rhyolitic volcanism of the western Cordillera in the border area of Chile and Bolivia consists of ignimbrite sheets, subvolcanic stocks, lavas and tuffs of dacite to rhyolite composition with initial Sr isotope ratios of 0.705-0.713 and ϵ_{Nd} -4 to -8 (Klerkx et al. 1977; Hawkesworth et al. 1982; Hildreth and Moorbath 1988). The volcanic rocks belong to the magnetite series (Fe_2O_3/FeO mostly >2) and are locally associated with polymetallic, Au-bearing epithermal systems (Cabello 1986; Gardeweg et al. 1984).

5. Rhyodacitic ignimbrites and rhyolitic to dacitic lavas and tuffs from the Altiplano and eastern Cordillera of Bolivia have Upper Cenozoic age and cover the same Sr and Nd isotope compositional field as equivalent rocks from the Chile-Bolivia border area given above (Klerkx et al. 1977; Schneider 1985; Redwood 1986). Subvolcanic stocks and brecciated vents are associated with polymetallic hydrothermal systems. Erosion level controls the style of mineralization, with tin-tungsten porphyry deposits at deeper levels (→ Llallagua) and epithermal precious metal-rich mineralization at shallow levels (Grant et al. 1977, 1980). Uranium deposits of roll-front type occur in the Altiplano (Michel and Schneider 1978).

Some geochemical data on the sample groups studied are compiled in Table 6. The Chilean rock samples were analyzed by atomic absorption spectrometry (detection limit: 0.2 ppm Sn), the remainder by X-ray fluorescence spectrometry (detection limit: 2-3 ppm Sn).

The tin data of the granitic samples are plotted in Figs. 75 and 76 as a function of Rb/Sr and TiO_2. Two distribution patterns can be distinguished: (1) Positive or negative linear correlation for log[Sn] vs. log[Rb/Sr] or log[Sn] vs. TiO_2 (wt%), respectively; (2) no or weak correlation of Sn with the two indicators of fractionation, i.e. constantly low Sn content for a large interval of Rb/Sr and TiO_2 values. The two distribution patterns characterize the tin granite population from northern Bolivia and the non-tin granite population from Chile and NW Argentina, respectively. The correlation lines for the tin granite suite are statistically significant at the 99.9 % level of certainty, with

---

Table 6 (next two pages). Chemical parameters of sample groups from the Central Andes on which tin data are available (arithmetic means ± 1 standard deviation). I and M means ilmenite- and magnetite-series rocks in the sense of Ishihara (1977)

Table 6 (continued)

Sample group	SiO$_2$ (wt%)	Rb (ppm)	Sr (ppm)	TiO$_2$ (wt%)	Rb/Sr	Sn (ppm)	Rock suite	References
North-Central Chile granites (Antofagasta and Copiapo transects)								
Biotite granites with SiO$_2$ >68 wt% (Late Paleozoic) (n = 16)	73.0 ±2.2	133 ±49	116 ±77	0.18 ±0.11	2.65 ±3.35	1.4 ±1.1	I/M	Ishihara et al. (1984) Shibata et al. (1984)
Biotite-hornblende granodiorites/quartz monzonites (SiO$_2$ <68 wt%) (Mesozoic-Cenozoic) (n = 18)	62.5 ±3.6	84 ±46	466 ±155	0.60 ±0.20	0.19 ±0.12	1.4 ±0.6	M	Ishihara et al. (1984) Shibata et al. (1984)
Northern Bolivian tin granites								
Huato biotite granite (Triassic) (n = 9)	68.8 ±0.9	236 ±14	149 ±9	0.50 ±0.06	1.59 ±0.12	12 ±8	I	Tistl (1985)
Illampu biotite-hornblende granodiorite (Triassic) (n = 9)	68.8 ±1.0	173 ±23	244 ±22	0.53 ±0.06	0.72 ±0.16	7 ±3	-	Miller (1988)
Zongo-Yani biotite-muscovite granite (Triassic) (n = 27)	73.2 ±1.4	374 ±137	86 ±62	0.20 ±0.11	5.91 ±3.59	21 ±11	I	Tistl (1985) Lehmann (1979)
Huayna Potosi biotite-muscovite granodiorite (Triassic) (n = 10)	69.1 ±2.9	215 ±95	210 ±95	0.47 ±0.22	2.50 ±3.21	13 ±12	I	Lehmann (1979)
Chacaltaya biotite-muscovite granite (Triassic) (n = 8)	71.3 ±2.3	448 ±140	56 ±24	0.22 ±0.12	11.4 ±10.2	29 ±9	I	Lehmann (1979)
Chacaltaya granite/greisen (tourmaline-muscovite alteration) (n = 16)	68.9 ±3.7	423 ±43	453 ±477	0.35 ±0.12	6.38 ±7.98	1112 ±2400	I	Lehmann (1979)
Unduavi biotite granodiorite (Triassic) (n = 10)	n.d.	177 ±20.3	366 ±20.9	0.599 ±0.047	0.488 ±0.070	4.9 ±2.0	I	Lehmann and Winkelmann (unpublished)
Taquesi biotite granodiorite/granite (Triassic) (n = 12)	69.5 ±3.2	217 ±56	258 ±101	0.40 ±0.17	1.12 ±0.82	17 ±4	I	Winkelmann (1983)
Chojlla muscovite granite (Triassic) (n = 7)	73.4 ±2.7	744 ±88	20 ±7	0.03 ±0.02	44.7 ±20.2	242 ±145	I	Winkelmann (1983)
Quimsa Cruz biotite granodiorite/granite (Tertiary) (n = 15)	68.6 ±3.7	303 ±93	352 ±251	0.53 ±0.21	2.13 ±3.10	13 ±10	-	Miller (1988)

Table 6 (continued)

Sample group	SiO₂ (wt%)	Rb (ppm)	Sr (ppm)	TiO₂ (wt%)	Rb/Sr	Sn (ppm)	Rock series	References
Pampean Ranges granites, NW Argentina								
Cafayate biotite granodiorite/tonalite (Ordovician) (n = 5)	64.1 ±2.6	79 ±25	275 ±54	0.46 ±0.14	0.32 ±0.15	<3.5	M	Rapela and Shaw (1979) Rapela et al. (1982)
Cafayate biotite-muscovite granite (Ordovician) (n = 18)	70.8 ±1.6	175 ±69	109 ±52	0.07 ±0.07	2.99 ±3.73	<3.5	I/M	Rapela and Shaw (1979) Rapela et al. (1982)
Cuchiyaco biotite-muscovite granodiorite (Ordovician) (n = 13)	69.0 ±1.2	95 ±10	151 ±26	0.14 ±0.05	0.65 ±0.20	<3.5	M	Rapela and Shaw (1979) Rapela et al. (1982)
Western Cordillera rhyolites								
Northern Chile and westernmost Bolivia (Cenozoic) (n = 39)	72.9 ±4.9	186 ±76	107 ±123	0.28 ±0.24	4.53 ±4.23	1.8 ±0.5	M	Klerkx et al. (1977) Pichler & Zeil (1969, 1972)
Guatiquina volcanic rocks, Chile								
Rhyolitic rocks (>68 wt% SiO₂) (Tertiary) (n = 11)	70.2 ±2.2	182 ±26	237 ±76	n.d.	0.91 ±0.45	2.2 ±0.8	M	Gardeweg et al. (1984) Hawkesworth et al. (1982)
Andesitic rocks (< 68 wt% SiO₂) (Tertiary) (n = 12)	62.5 ±3.0	139 ±24	352 ±108	n.d.	0.44 ±0.15	2.0 ±0.5	M	Gardeweg et al. (1984) Hawkesworth et al. (1982)
Volcanic rocks of the Bolivian Altiplano								
Southern Altiplano rhyodacites (Cenozoic) (n = 54)	65.3 ±9.5	170 ±62	281 ±79	0.65 ±0.17	0.66 ±0.38	3.0 ±0.5	M	Kussmaul et al. (1977) Fernandez et al. (1973
Chorolque rhyodacite (tourmaline-sericite alteration) (n = 22)	n.d.	276 ±57	147 ±116	0.62 ±0.03	4.77 ±7.61	75 ±139	-	Lehmann and Sanchez (unpublished)
Los Frailes Ignimbrite (rhyodacite) (Tertiary) (n = 19)	65.7 ±3.1	233 ±27	408 ±76	0.70 ±0.15	0.59 ±0.10	4.5 ±0.7	M	Michel (in preparation) Schneider (1985)
Northern Altiplano rhyodacites (Cenozoic) (n = 7)	66.6 ±1.7	132 ±21	711 ±213	0.66 ±0.09	0.20 ±0.05	<3.5	-	Redwood (1986)

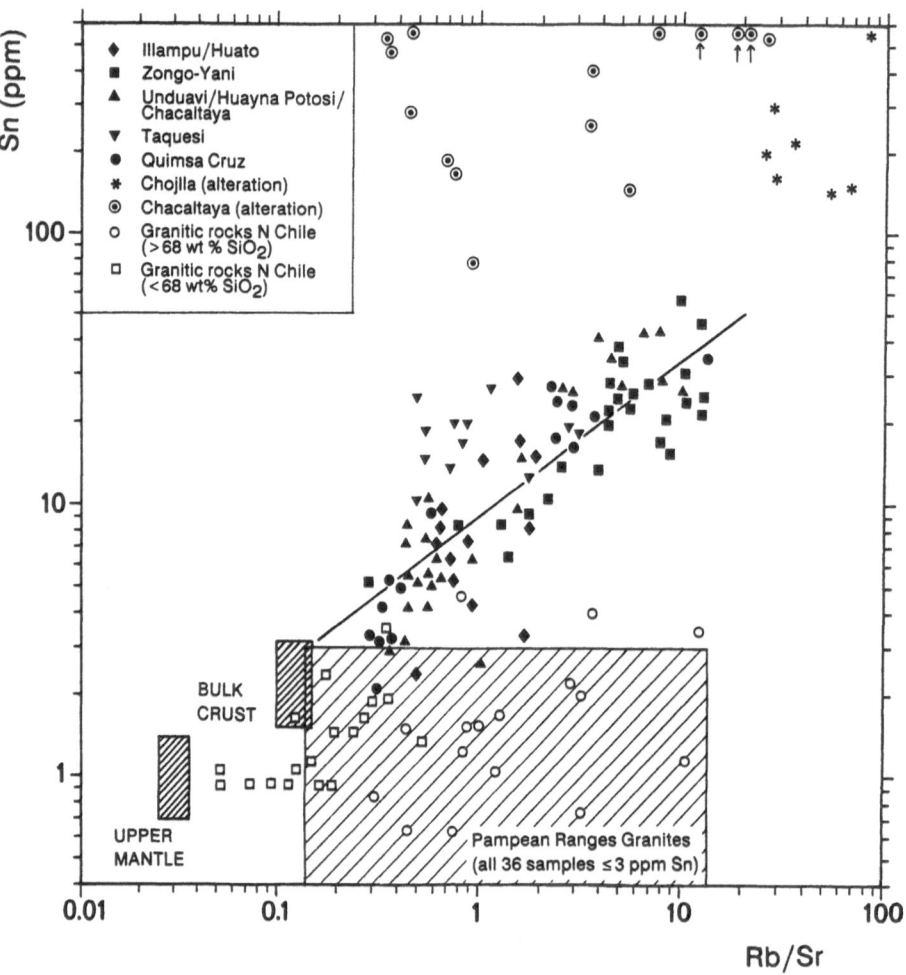

Fig. 75. Rb/Sr-Sn variation diagram for granitic rocks of the Central Andes. Tin granites of the northern Bolivian tin belt define a correlation line log[Sn] = 0.956 + 0.565log[Rb/Sr] with r = 0.78 (n = 98). Hydrothermally overprinted granite samples from the Chacaltaya and Chojlla tin ore systems give scatter distributions at high tin levels. Non-tin granites of northern Chile and NW Argentina (Pampean Ranges) have low tin contents and give no systematic tin enrichment trend during magmatic evolution. Bulk Crust and Upper Mantle reference fields according to Anderson (1983) and Taylor and McLennan (1985)

correlation coefficients of r(logRb/Sr-logSn) 0.78 and r(TiO2-logSn) -0.74 (n = 98). The geometric parameters of the correlation lines are very similar to those of tin granite suites in other parts of the world (Lehmann 1982).

Fractional crystallization is the dominant petrogenetic process which controls the magmatic evolution of both the Cordillera Real/Quimsa Cruz tin granites

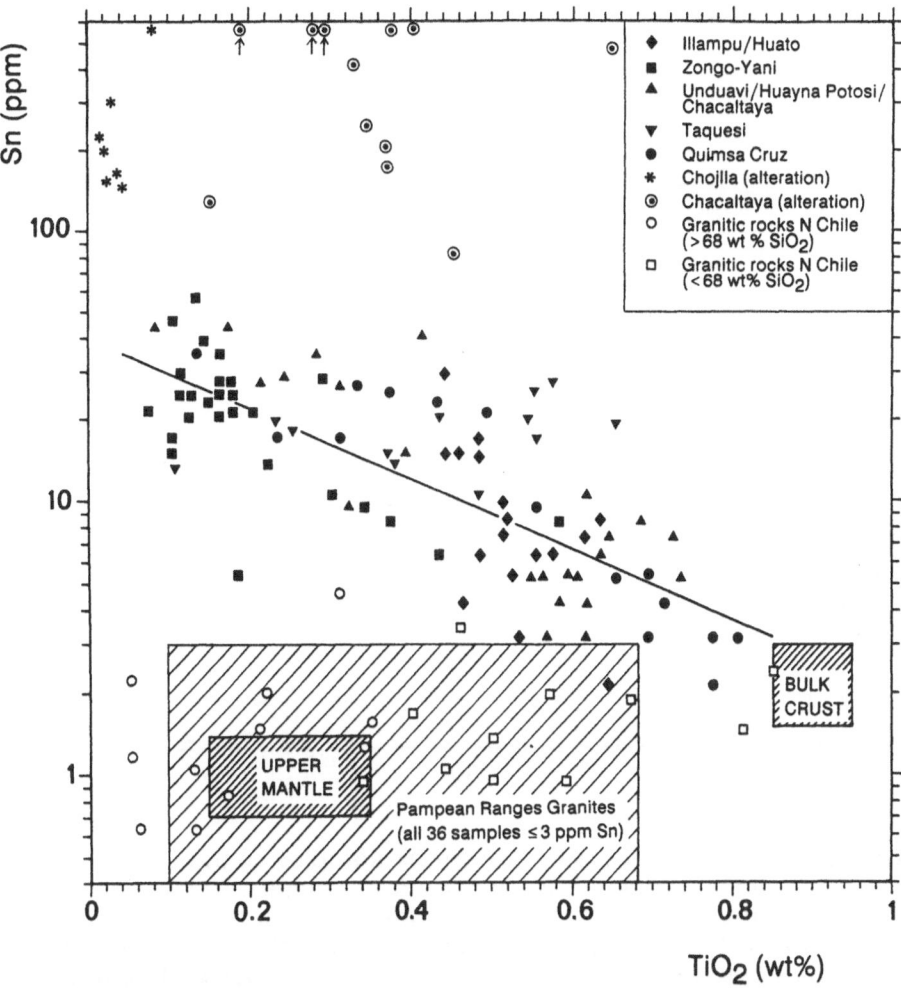

Fig. 76. TiO$_2$-Sn variation diagram for granitic rocks of the Central Andes. Tin granites of the northern Bolivian tin belt define a correlation line log[Sn] = 1.581 - 1.252[TiO$_2$] with r = -0.74 (n = 96; hydrothermally overprinted sample populations from Chojlla and Chacaltaya mining areas are not included)

(Lehmann 1979, 1985) and the non-tin granites of NW Argentina (Rapela and Shaw 1979; Saavedra et al. 1987). The Rb-Sr plot in Fig. 77 underlines these earlier conclusions. Slope and length of the correlation trends point to the importance of plagioclase fractionation in both granite populations. The two different tin distribution patterns can therefore be interpreted to result from two different tin distribution coefficients, with $\bar{D}_{Sn}$(xtls/melt) ≈ 1 in the non-tin granites and $\bar{D}_{Sn}$(xtls/melt) < 1 in the tin granites (Lehmann 1982). The correlation trends for the tin granite suite in Figs. 75 and 76 extend to the bulk-crust reference field, least evolved granite samples plot close to average

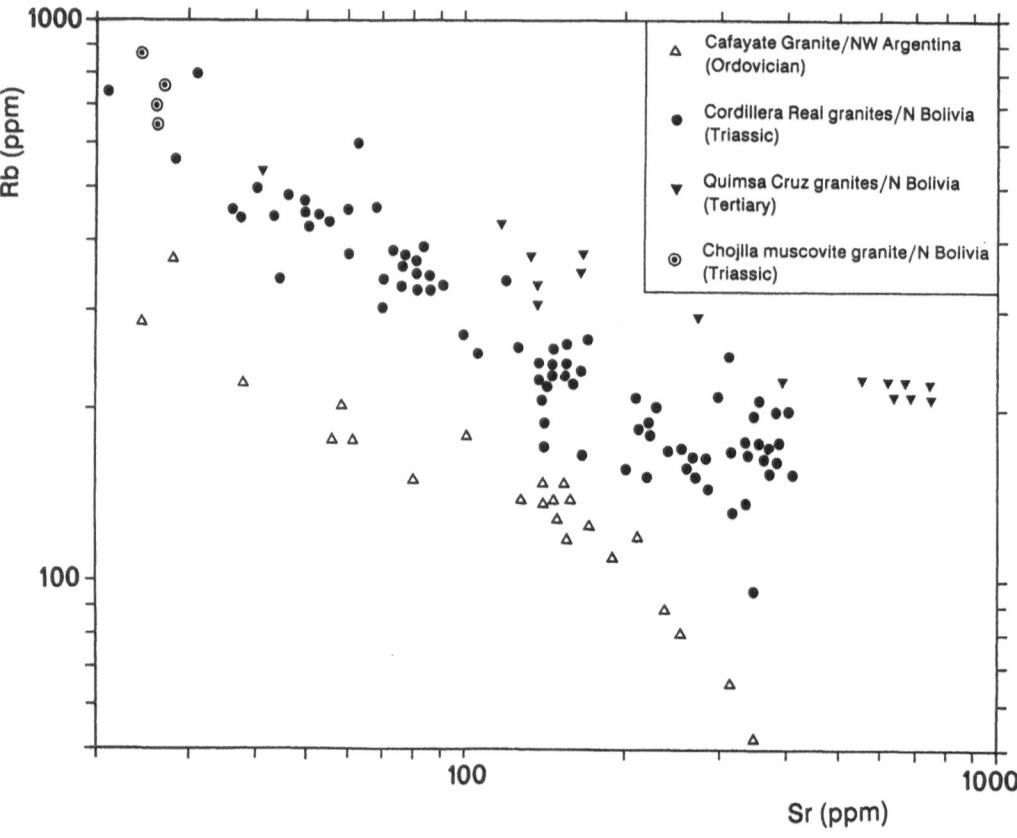

Fig. 77. Rb-Sr variation diagram for rock samples of the tin-barren Cafayate granite (Pampean Ranges, NW Argentina), the Cordillera Real tin granites, the Quimsa Cruz tin granites, and the muscovite granite of the Chojlla tin mine. Slope and extent of the linear correlation trends in log-log space suggest substantial plagioclase fractionation in all sample suites

crustal material. This situation points to an origin of the tin granites from a melt initially not anomalous in tin, and to source rocks of average crustal composition. A crustal source of the Bolivian tin granites is likely in view of their peraluminous S-type character (Clark and Robertson 1978; Lehmann 1979; Winkelmann 1983; Tistl 1985) and of their isotope characteristics (Kontak et al. 1984; McNutt and Clark 1983; Miller and Harris 1989).

The low initial Sr-isotope data of the granitic rocks from northern Chile and NW Argentina, on the other hand, suggest an important contribution by mantle material (Shibata et al. 1984; Rapela et al. 1982; Saavedra et al. 1987). Tin contents in these sample groups are very low (Chile: 1.4 ppm Sn; Argentina: all 36 samples analyzed <3.5 ppm Sn), in between average tin contents for bulk crust and upper mantle. These non-tin granite suites are predominantly of magnetite-series affiliation with Fe_2O_3/FeO ratios of >0.5, distinctly different from the Bolivian ilmenite-series tin granites with Fe_2O_3/FeO ratios of <0.1 (Lehmann unpubl.). Both Fe^{3+}/Fe^{2+} ratio and opaque mineral assemblage reflect the oxygen fugacity during rock formation, which in magnetite-series granitic rocks is relatively high and favours the accessory mineral assemblage of magnetite-titanite (Ishihara 1981; see Chap. 2.4). Magnetite and titanite have exceptionally high tin distribution coefficients of $D_{titanite}(xtl/melt) \approx 60$ and $D_{magnetite}(xtl/melt)$ 4-12 in granitic melts (Antipin et al. 1981), which may provide an explanation for the bulk tin distribution coefficient in magnetite-series rocks close to unity.

The hydrothermally overprinted granite samples of the Chacaltaya and Chojlla tin-tungsten mines in Bolivia give scatter patterns in Figs. 75 and 76. The Chacaltaya greisen samples are characterized by feldspar-destructive alteration and have the stable mineral assemblage quartz-muscovite-tourmaline-apatite-siderite-fluorite-cassiterite. The Chojlla aplogranite has the stable assemblage quartz-oligoclase-microcline together with sub-solidus muscovite-tourmaline-apatite-cassiterite. Both the Chacaltaya and Chojlla samples show element distribution patterns established by fractional crystallization and enhanced and modified by fluid interaction.

The tin data of the volcanic rocks are plotted in Fig. 78 as a function of Rb/Sr. Titanium or zirconium contents are not very useful as indicators of fractionation in this rock group, which in the Bolivian sector has mostly SiO_2 ≤ 66 wt%, because these components become less compatible towards intermediate melt composition and higher melt temperature, i.e. change their bulk distribution coefficients significantly.

The tin distribution pattern of the volcanic rocks is similar to the one in the granites: weak or no tin enrichment during magmatic evolution in barren volcanics, and distinct tin enrichment in samples from the Bolivian tin belt, where the least-evolved samples are near the bulk-crust reference field. The hydrothermally overprinted rhyodacitic rocks of the Chorolque tin porphyry system (tourmaline-sericite alteration) are characterized by high Rb/Sr ratios due to plagioclase destruction and Sr leaching. This fluid overprint is accompanied by tin enrichment up to subeconomic levels. There is a tendency to chemical continuum between magmatic and hydrothermal processes, which complicates the distinction between both stages.

Data for a classification of the Bolivian tin-bearing volcanics into ilmenite- or magnetite-series rocks are not sufficient and difficult to obtain in view of

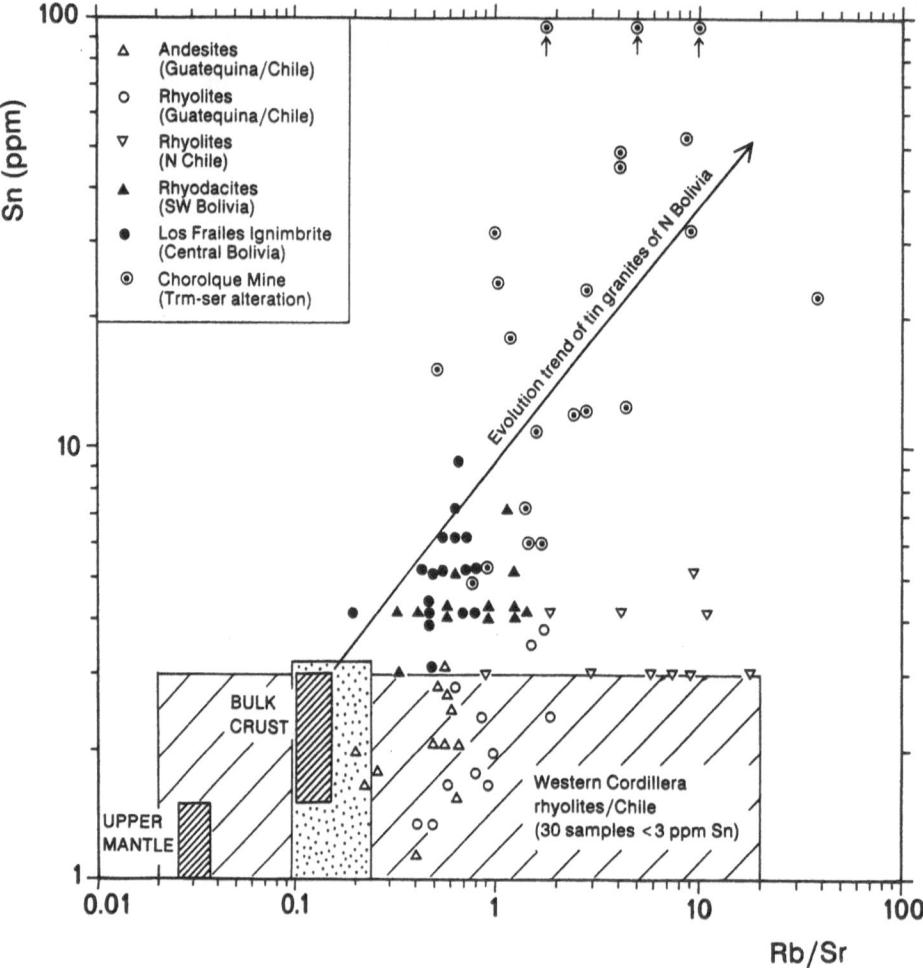

Fig. 78. Rb/Sr-Sn variation diagram for volcanic rocks of the Central Andes. The reference line gives the evolution trend of the tin granites of the northern Bolivian tin belt defined in Fig. 75. Rhyolitic rocks from the western Cordillera (Chile and westernmost Bolivia) have low tin contents and give no or very little tin enrichment during magmatic evolution. Rhyodacitic rocks from the Los Frailes Ignimbrite in Central Bolivia plot close to least-evolved tin granite samples of the Cordillera Real. Hydrothermally overprinted rocks from the Chorolque tin porphyry system give a scatter distribution at high tin levels. The stippled pattern enclosing the Bulk Crust reference field locates northern Altiplano rhyodacites (7 samples with ≤3 ppm Sn; Redwood 1986)

widespread alteration. Hydrothermal overprint and mineralization developed under conditions of relatively low oxygen fugacity in between the NNO and

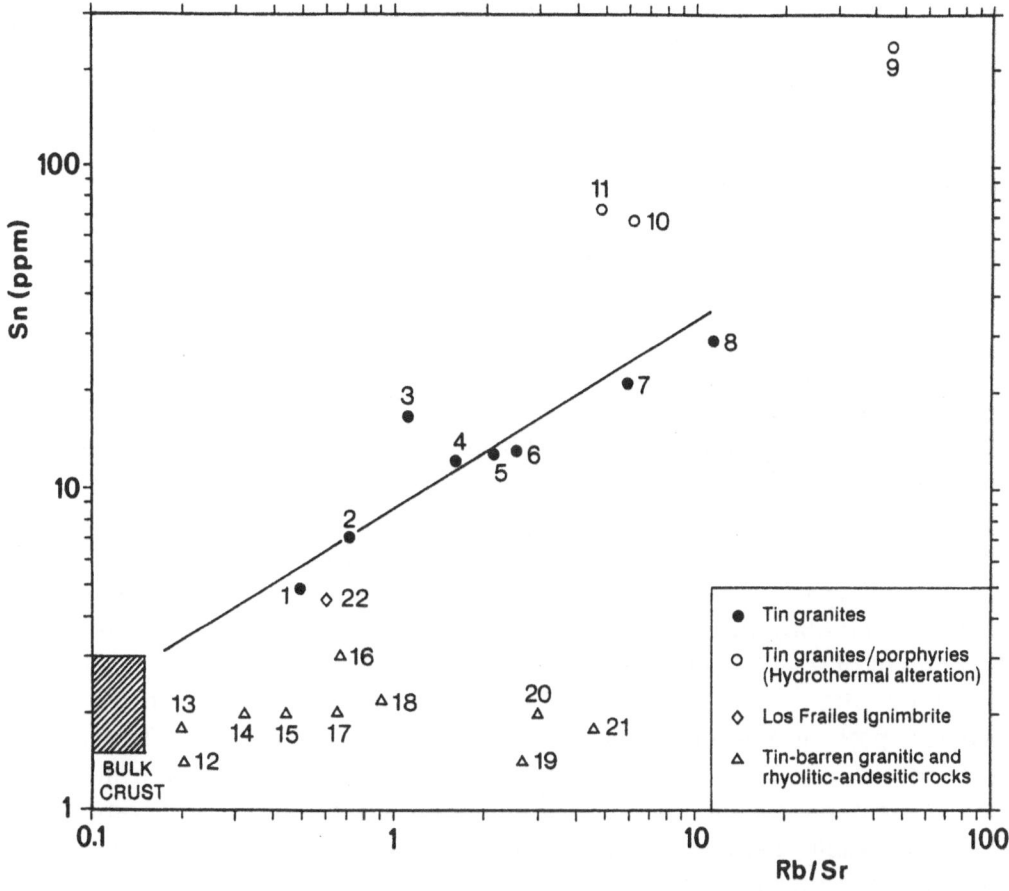

Fig. 79. Synoptic Rb/Sr-Sn variation diagram for arithmetic means of the sample populations studied. Tin-bearing granites define a tin enrichment trend as a function of degree of magmatic evolution. This trend extends back to average crustal rock composition. The Los Frailes ignimbrite from the periphery of the Central Bolivian tin belt plots close to the least-evolved part of this trend. Hydrothermally modified samples from tin ore systems have high tin levels, partly they are also most evolved magmatically (compare TiO_2 contents in Fig. 76 and Table 6). Tin barren granitic and rhyolitic rocks have low tin contents which do not correlate with degree of magmatic fractionation. 1 Unduavi granodiorite; 2 Illampu granodiorite; 3 Taquesi granodiorite/granite; 4 Huato granite; 5 Quimsa Cruz granites; 6 Huayna Potosi granite; 7 Zongo-Yani granite; 8 Chacaltaya granite; 9 Chojlla granite; 10 Chacaltaya greisen; 11 Chorolque rhyodacite; 12 N Chile grano-diorites; 13 Northern Altiplano rhyodacites; 14 Cafayate granodiorite; 15 Guatiquina andesites; 16 SW Bolivian rhyodacites; 17 Cuchiyaco granodiorite; 18 Guatiquina rhyolites; 19 N Chile granites; 20 Cafayate granite; 21 Western Cordillera rhyolites; 22 Los Frailes ignimbrite

QFM buffers (as inferred from the general paragenetic scheme of Kelly and Turneaure 1970), with fluids in an early stage of magmatic origin (Grant et al. 1980). Given a rock-buffered fluid evolution an ilmenite-series affiliation of the volcanics seems probable. This contrasts with the rhyolites of the western Cordillera which belong to the magnetite series. The source material of both tin-bearing volcanics of the eastern Cordillera and tin-barren volcanics of the western Cordillera appears to be the lower crust (Sr and Nd isotope data by Klerkx et al. 1977; Hawkesworth et al. 1982; Schneider 1985; Redwood 1986; Hildreth and Moorbath 1988).

It is interesting to note that the Miocene-Pliocene Macusani ash-flow tuffs in SE Peru, about 60 km west of the San Rafael Sn-Cu deposit, have chemical characteristics very similar to the highly evolved portions of the tin granites of northern Bolivia. The Macusani volcanics have been studied by Noble et al. (1984) and Pichavant et al. (1987, 1988a,b) and have been discussed in petrogenetic terms of fractional crystallization and partial melting, the latter process being favoured by Pichavant et al. (1988b). The chemical data from Pichavant (1988b), thought to reflect the magmatic chemistry, give a range in tin content of 27-194 ppm (nine samples) with corresponding TiO_2 values of 0.22-0.04 wt%, and Rb/Sr of 3.4-710. These data would plot in the most evolved part of the correlation trends of Figs. 75 and 76. The highly fractionated nature of the Macusani volcanics is also displayed in systematic and characteristic enrichment and depletion patterns for other incompatible and compatible elements. These features, together with the ilmenite-series nature of the rocks (Noble et al. 1984; Pichavant et al. 1988a), identify the Macusani volcanic zone as a very promising area for tin mineralization in a porphyry environment.

In summary, it can be concluded that the geochemically better-defined Bolivian tin granites have systematic tin enrichment trends as a function of degree of fractionation. The non-tin granites from the periphery of the Bolivian tin belt show no analogous tin enrichment which can be understood as a consequence of an unfavourable bulk tin distribution coefficient near unity (magnetite-series granitic rocks). The least fractionated parts of the tin granite suite have tin contents typical of average crust, which does not support the assumption of anomalous tin contents in the source material of the Bolivian tin granites (Fig. 79).

5.4 Kelapa Kampit, Belitung Island, Indonesia

The Kelapa Kampit tin deposit on Belitung Island is the economically most important example of strata-bound tin mineralization in the SE Asian tin belt. It has often been cited as an example of synsedimentary precursor mineralization later remobilized during granite magmatism, and is therefore of great importance in a general metallogenic context. Kelapa Kampit is also a key locality for the concept of lithophile-element mineralization associated with mafic volcanism, which was first put forward for tungsten by Maucher (1965) and later extended to tin mineralization by Plimer (1980). The genetic interpretation of an ore deposit is in this concept primarily based on geometric parameters which have been emphasized by Amstutz (1959:12): "Congruency may be used as an indication of syngenetic origin". This claim invites confusion between the phenomenology and nature of an ore deposit, because hydrothermal fluid circulation and ore deposition are controlled by the permeability structure of the enclosing medium, which in turn depends on lithological parameters. Inherent lithological contrasts will result in variable degree of stratabound fluid flow at a scale defined by the local geology.

The Kelapa Kampit Mine is today focussed on open-pit bulk mining of the Nam Salu horizon. Earlier selective mining of a large cassiterite-sulphide vein system (bedding-plane lodes and discordant veins) has been abandoned. The Nam Salu horizon is a steeply dipping, 15-40-m- thick, iron-rich phyllite which consists of laminated pyrrhotite-pyrite-magnetite in a matrix of stilpnomelane-biotite-chlorite, with some actinolite-rich calcsilicate lenses. Cassiterite occurs in fine-grained disseminations (average grain size around 50 μ) over a strike length of 3 km at a geochemical level of 200 ppm Sn, and in the Nam Salu open pit over a length of about 100 m reaches an economic grade of 1-1.5 wt% Sn (Greg Bolton, PT Preussag, pers. commun. 1985). The Nam Salu horizon is part of a Permo-Carboniferous low-grade rock sequence which consists of predominantly clastic sediments with intercalations of limestone, chert, and ironstone associated with submarine mafic volcanism (Adam 1960; Osberger 1962). The sequence is intruded by Triassic granites which are tin-bearing (see Tanjungpandan Pluton; Chap. 4.1). Igneous rocks in the immediate Kelapa Kampit area are however restricted to rare porphyry dikes; larger granite outcrops are 20 km away.

The Nam Salu ore horizon has relic tuff fabrics and was identified by Schwartz and Surjono (1986) as metasomatically altered, tuffitic banded iron formation. The tin-mineralized rock portions have high contents of Rb, Cs, Be, Pb, Bi, W and Zn; a feature distinctly different from sea-water overprinted meta-basalts

Table 7. Arithmetic means of some element contents in major rock units of the Kelapa Kampit mining area, Indonesia. Data from Schwartz and Surjono (1986), meta-basalt reference data from Lehmann (1988b).

	1	2	3	4	5	6	7
SiO_2	34.23	32.49	53.68	82.19	90.33	46.03	48.20
TiO_2	1.08	1.46	1.21	0.35	0.14	0.72	3.27
Al_2O_3	9.62	11.61	12.80	6.36	3.25	6.36	12.95
Fe_2O_3	39.87	40.72	19.42	5.26	2.66	27.12	14.42
MnO	0.38	2.14	0.37	0.15	0.10	0.66	0.18
MgO	3.64	3.20	3.36	1.16	0.81	3.28	5.55
CaO	0.13	0.43	1.00	0.02	0.01	11.62	8.83
Na_2O	0.07	0.09	0.57	0.16	0.13	0.28	2.07
K_2O	4.16	2.33	2.28	1.37	0.79	0.31	0.68
P_2O_5	0.10	0.09	0.12	0.03	0.02	0.10	0.40
L.O.I.	3.77	3.83	3.46	2.10	1.18	2.34	3.05
Total	97.05	98.38	98.28	99.16	99.40	98.81	99.60
n	23	7	5	19	2	3	6
Ba	403	370	114	178	123	152	18
Bi	163	<6	<6	<6	<6	<6	<6
Ce	69	80	81	62	37	37	96
Cr	116	115	86	43	76	46	74
Cs	325	-	-	-	-	-	-
Cu	169	17	31	32	36	198	114
F	603	-	-	-	-	-	542
La	155	133	70	34	13	37	176
Nb	52	13	12	7	5	11	13
Ni	41	41	41	17	13	91	28
Pb	216	4960	3000	1200	447	<5	<5
Rb	1380	577	317	129	81	39	19
Sc	13	22	19	5	4	16	31
Sn	16400	90	42	127	135	434	<3
Sr	13	20	54	8	4	100	201
Ta	<5	<5	<5	<5	<5	<5	<5
Th	30	26	24	24	21	16	12
U	13	6	4	3	<3	6	4
V	152	215	151	83	26	116	295
W	115	12	21	9	<5	<5	<5
Y	89	33	33	18	7	24	44
Zn	354	3630	5030	1050	293	788	115
Zr	78	106	32	152	30	63	204

1 = Nam Salu horizon in open pit (stilpnomelane-biotite-chlorite-magnetite-pyrrhotite-pyrite phyllite)
2 = Nam Salu horizon in exploration drilling (core samples)
3 = Country rock of Nam Salu horizon (chlorite schist)
4 = Country rock of Nam Salu horizon (meta-siltstone)
5 = Country rock of Nam Salu horizon (chert)
6 = Country rock of Nam Salu horizon (calcsilicate rock)
7 = Meta-basalt (spilite), 35 km NW of Kelapa Kampit

n = number of samples analyzed. Fe_2O_3 is total iron.

(spilites) from the same sequence several kilometers away, which have Cr, Ni and V levels similar to the Nam Salu horizon without, however, any lithophile element enrichments (Table 7). The high Rb and Cs contents of the Nam Salu horizon argue against a sea water-dominated hydrothermal system. The mineralogical-geochemical character of the mineralization suggests a granite-related situation in the sense of sulphide replacement, with the reactive material consisting of mafic tuff which is petrochemically very close to dolomitic carbonate material. The potassic alteration style and the crystal morphology of the Nam Salu cassiterite (short prismatic habitus) point to relatively high temperature of the hydrothermal overprint/mineralization (Omer-Cooper et al. 1974; Ahlfeld 1931; Kelly and Turneaure 1970; Rutherford 1969).

The conventional model which infers a metal source by means of an unexposed tin granite has yet to be positively proven by drilling or radiometric dating of the mineralizing event. There is the analogous case of the Sangdong scheelite deposit in Korea which was resolved recently. The Sangdong strata-bound tungsten mineralization in Cambrian limestones indicated, through its mineralogical, geochemical and stable isotope zoning pattern, a granitic intrusion at depth. Recent drilling revealed 500 m beneath the main orebody the expected Upper Cretaceous highly evolved granite (Moon 1988, 1989).

6 Model of Tin Ore Formation

6.1 Origin of Tin Granites

The origin of tin granites is part of the ongoing discussion on the origin of granitic rocks in general. There is an historic polarization in petrogenetic models for granites which are seen either in idealized terms of intracrustal melting or of differentiation of mantle melts. More recent chemical and isotopic data suggest a multiple-source and open-system model for the formation of granitic rocks, in which continental silicic magmatism is seen as a result of variable physical and chemical interaction of upper mantle and continental crust, i.e. crust-mantle mixtures rather than either pure differentiates of mantle magmas or pure products of crustal anatexis (Hildreth 1981; Holden et al. 1987; Marsh 1987; DePaolo 1988). The state of crustal stress will critically control the process time for chemical interaction between ponded melts and their host rocks in a lower crustal environment, and together with crustal thickness (length of percolation column) will also define the process time for the internal evolution of silicic melts during their rise to upper crustal levels (Pitcher 1987; Hildreth and Moorbath 1988; Lipman 1988). Crustal melting and mixing of crustal and subcrustal melt portions play its largest role in the high-temperature deeper (lower crustal) levels of magmatic systems. At shallower levels, assimilation is minor and crystal fractionation appears to be the dominant process controlling the magmatic evolution of high-silica melts (Farmer and DePaolo 1983, 1984; Musselwhite et al. 1989).

The fundamentally continuous nature of granite petrogenesis has been demonstrated with the help of Sr and Nd isotope data for the Mesozoic and Tertiary granitic rocks of the southwestern USA (Farmer and DePaolo 1983, 1984), for the Paleozoic batholiths of southeastern Australia (McCulloch and Chappell 1982), and for the Paleozoic to Cenozoic granite belts along the western margin of South America (Pankhurst et al. 1988). These granite provinces cover in terms of the petrogenetic parameters of initial ϵ_{Sr}-ϵ_{Nd} a broad concave compositional field, defined by a negative correlation trend for values of ϵ_{Sr} up to 100, and relatively constant ϵ_{Nd} values for a wide range of ϵ_{Sr} of >100.

Figure 80 is a compilation of the available initial Sr and Nd data on tin granites. The data cover a wide compositional range similar to the variation in other tin-

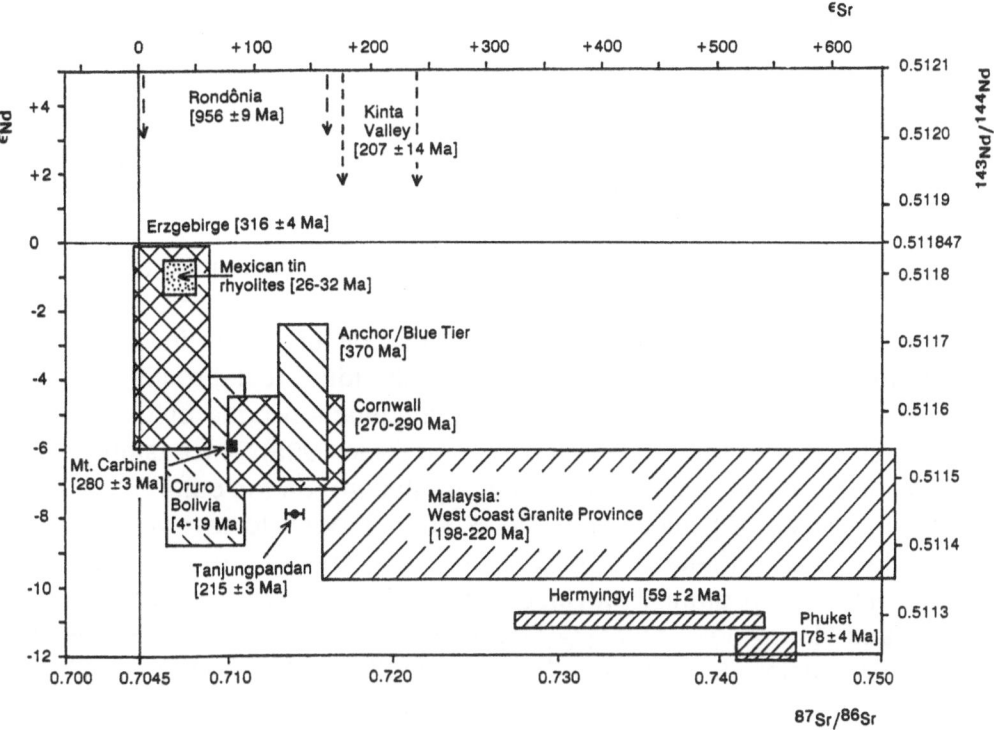

Fig. 80. Initial Sr and Nd isotope data for granitic rocks from several tin
provinces. Data sources: Phuket (Cobbing written commun. 1989;
Putthapiban et al. 1986); Hermyingyi, Burma (Cobbing written
commun. 1989; Darbyshire and Swainbank 1988); Tanjungpandan,
Indonesia (Darbyshire 1988b; Darbyshire pers. commun. 1988); East
and West Coast granites, Malaysia (Liew and McCulloch 1985); Mt.
Carbine, North Queensland, and Anchor Mine, Tasmania, Australia
(Higgins and Sun 1988); Cornwall (Darbyshire and Shepherd 1988);
Oruro, Bolivia (Redwood 1986); Erzgebirge, Germany (Gerstenberger
1989); Sierra Madre Occidental, Mexico (Ruiz 1988); Cordillera Real
(McNutt and Clark 1983; Miller and Harris 1989); Kinta Valley,
Malaysia (Schwartz and Askury 1989); Rondônia, Brazil (Priem et al.
1989).

barren granite provinces (cf. DePaolo 1988), and do not suggest any specific
origin for tin granites which would set them apart from non-tin granites. The
discussion of these data must take into account that highly fractionated melts
have low Sr and high Nd concentrations and are thus susceptible, even with
small amounts of assimilation, to large shifts in $^{87}Sr/^{86}Sr$ but not in
$^{143}Nd/^{144}Nd$. The same limitation applies also to Pb isotope systematics of
contaminated mantle melts which are readily equilibrated with external Pb by
minor crustal input. The Sr system in late, highly fractionated melt phases

may also shift towards higher Sr initials as a consequence of the internal Rb-Sr evolution in large melt reservoirs (McCarthy and Cawthorn 1980). On the other hand, the postmagmatic rubidium metasomatism important in some tin granites can result in too low apparent Sr initial ratios (Gerstenberger 1989). The importance of mantle-derived components in granitic rocks can be adequately assessed only by Nd isotope data.

The source material of the Malaysian Main Range granites of the SE Asian tin belt with $\epsilon_{Nd}(T)$ of -6 to -10 must be predominantly Phanerozoic crust with a Nd model age of 1300-1700 Ma (Liew and McCulloch 1985). Involvement of mantle material would require mixing with still older crustal rocks, which is, however, not likely in view of the upper intersection ages of U-Pb zircon reverse discordia of 1500-1700 Ma (Liew and Page 1985). The eastern granite province in SE Asia (East Coast granite population), on the other hand, has a large spread in $\epsilon_{Nd}(T)$ values of -0.7 to -6.2 and appears to have variable input of mantle material (Liew and McCulloch 1985; Darbyshire pers. commun. 1988). The western granite province of the SE Asian tin belt (samples from Phuket and Hermyingyi) has $\epsilon_{Nd}(T)$ values of -11 to -12, which suggests a dominantly crustal source at least 1800 Ma old (Darbyshire unpubl. data).

The Nd and Sr isotope data from tin granites in Bolivia (Cordillera Real), Tasmania (Blue Tier batholith), North Queensland (Mt. Carbine) and of the Erzgebirge, as well as from tin porphyries in Bolivia and Mexico, suggest a variable degree of involvement of mantle material. This is particularly likely for the Erzgebirge tin granites which have $\epsilon_{Nd}(T)$ values of 0.0 to -6.0 (Gerstenberger 1989), and for the Mexican tin rhyolites with $\epsilon_{Nd}(T)$ around -1 (Ruiz 1988).

The data in Fig. 80 do not allow assigning an exclusively crustal origin to all tin granites, although crustal material is probably the dominant source of tin granites. This is true, however, for most tin-barren granites as well. Continental setting is a worldwide feature of tin-tungsten deposits and their associated igneous rocks. The tectogenetic environments of tin granites and porphyries are:
1. Postorogenic magmatism in continental collision belts, i.e. Permo-Triassic tin belt in SE Asia (Mitchell 1977; Beckinsale 1979), Hercynian tin provinces in western and central Europe (Mitchell 1974; Holder and Leveridge 1986).
2. Internal zones of active continental margins (back-arc regions), i.e. Tertiary tin porphyries in Bolivia, Mexico, eastern Siberia, USSR (Sillitoe 1976), Cretaceous-Tertiary tin granites in Thailand and Burma (Beckinsale 1979).
3. Intracontinental rift zones, i.e. Triassic tin granites in northern Bolivia (Kontak et al. 1984), Cretaceous tin granites in southern China (Chen

Guoda 1989), Cretaceous tin granites in Nigeria (Bowden and Kinnaird 1984), Precambrian tin granites in Rondônia, Brazil (Sillitoe 1974), Precambrian Bushveld granites (Hunter 1973).

The tectonic setting of tin deposits may reflect a critical physical parameter for tin formation, i.e. a thick crust which permits process space and time for a high degree of partial melting and extended fractional crystallization. More than 25 % partial melt during crustal anatexis is probably a precondition for large-scale magma segregation. A smaller amount of silicic melt is likely to remain trapped in-situ as migmatite (Thompson and Connolly 1990). Magma segregation and intrusion into an upper crustal level is favoured by brittle lithosphere in crustal extensional zones.

6.2 Time-Space Framework

The spatial association of tin ore deposits with granitic rocks is documented by a very large observational basis and is generally accepted. However, all genetically more specific interpretations meet less unanimous consent. The postmagmatic-epigenetic nature of hydrothermal tin deposits is a consequence of their geometry and age relative to the host granite. Radiometric age data commonly give a temporal hiatus between solidification of the granitic country rock and hydrothermal ore formation which can be on the order of several million years. The Cornwall situation is summed up in the Rb-Sr isochron data of Darbyshire and Shepherd (1985) in Table 8.

Table 8. Rb-Sr isochron data for granitic rocks and hydrothermal fluids from tin deposits (fluid inclusions in vein quartz) in Cornwall. Data from Darbyshire and Sheppard (1985)

Locality	Age	$^{87}Sr/^{86}Sr_i$
Carnmenellis pluton	290 ± 2 Ma	0.713 ± 4
Bodmin Moor pluton	287 ± 2 Ma	0.717 ± 2
St. Austell pluton	285 ± 4 Ma	0.7095 ± 9
Dartmoor pluton	280 ± 1 Ma	0.7094 ± 3
Meldon Elvan (aplite)	279 ± 2 Ma	0.710 ± 2
Wherry Elvan (aplite)	282 ± 6 Ma	0.712 ± 3
Brannel Elvan (aplite)	270 ± 9 Ma	0.715 ± 3
South Crofty mine (quartz)	269 ± 4 Ma	0.7147 ± 3
Geevor mine (quartz)	270 ± 15 Ma	0.712 ± 1

150

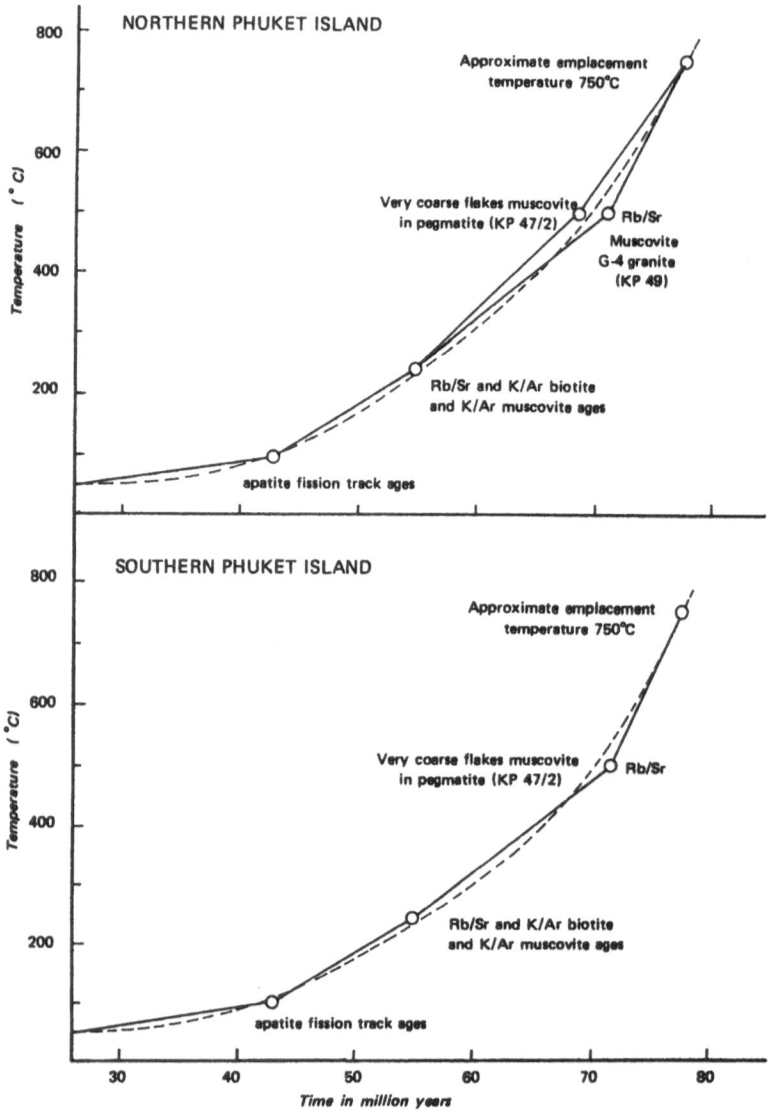

Fig. 81. Cooling history of the Phuket tin granites (Putthapiban et al. 1986:62). Emplacement age of granites: 78 ± 4 Ma (Rb-Sr isochron with inital $^{87}Sr/^{86}Sr$ 0.7428 ± 18)

K-Ar data suggest continued hydrothermal activity associated with the Cornwall granites into the Mesozoic (Halliday 1980; Jackson et al. 1982; Stone and Exley 1986). Long-lived fluid circulation induced by radiogenic heat production in highly fractionated granitic rocks with >8-10 ppm U ("high heat production granites"; Fehn et al. 1978; Darnley 1986) is generally believed to be responsible for intermediate- and low-temperature processes with a large spectrum of metal concentrations (F, Ba, U, Pb, Zn, Ag, Au, Sb, Co, Bi) of epithermal formation (300-50 $^\circ$C). The internal radiogenic energy released in "high heat production granites" seems, however, to be insufficient for any substantial redistribution of tin, as noted by Fehn (1985) for the Cornwall province.

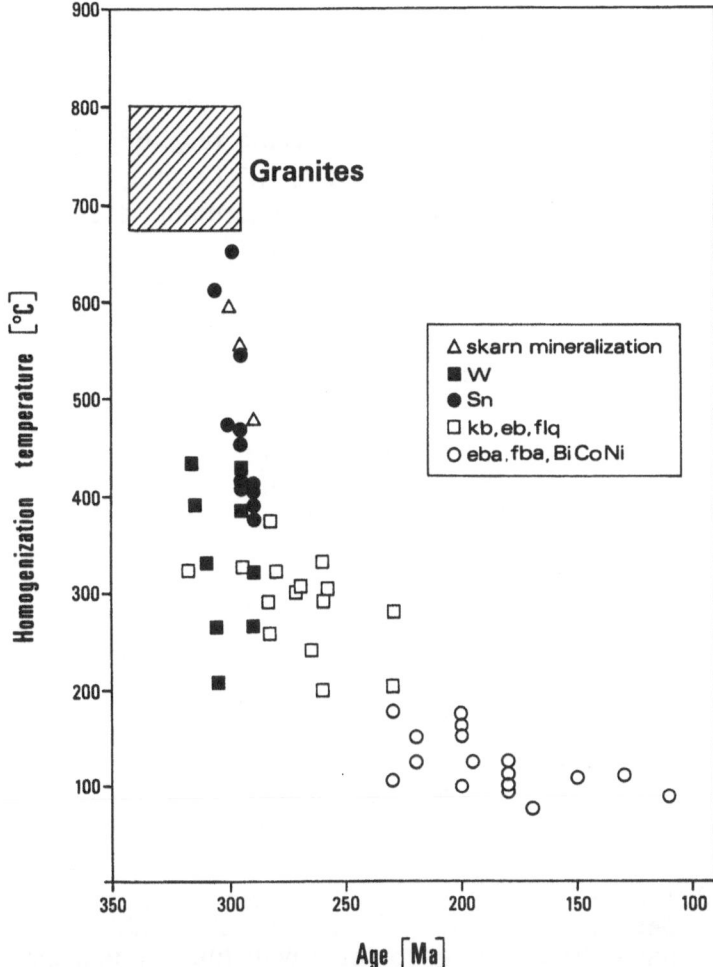

Fig. 82. Cooling history of the Erzgebirge granites according to radiometric and fluid-inclusion data from hydrothermal mineral assemblages of the early Sn-W mineralization and subsequent epithermal F-Ba-Ag-Pb-Zn-Bi-Co-Ni-U mineralization stages. (Thomas and Tischendorf 1987:31)

A similarly extended thermal (and hydrothermal) history has been documented for the Phuket granites in southern Thailand by a combination of Rb-Sr, K-Ar and fission track ages (Fig. 81) (Putthapiban et al. 1986). In the

Erzgebirge tin province, numerous U-Pb, Rb-Sr and K-Ar age data together with homogenization temperatures of fluid inclusions define a hydrothermal cycle lasting over 150 Ma (Fig. 82) (Thomas and Leeder 1986; Thomas and Tischendorf 1987).

The temperature-time diagrams demonstrate that tin granite provinces are characterized by persistent heat anomalies and have therefore long-lasting hydrothermal activity. The time period of tin mineralization can be of the order of several million years as shown by high-resolution $^{40}Ar/^{39}Ar$ age data from the Panasqueira Mine, Portugal, which give a duration of cassiterite deposition of greater than 4 million years (Fig. 83) (Snee et al. 1988). The multiphase nature and extended life span of both granite magmatism and tin mineralization require a composite magmatic system, large-scale fluid circulation and a long-lived fluid reservoir.

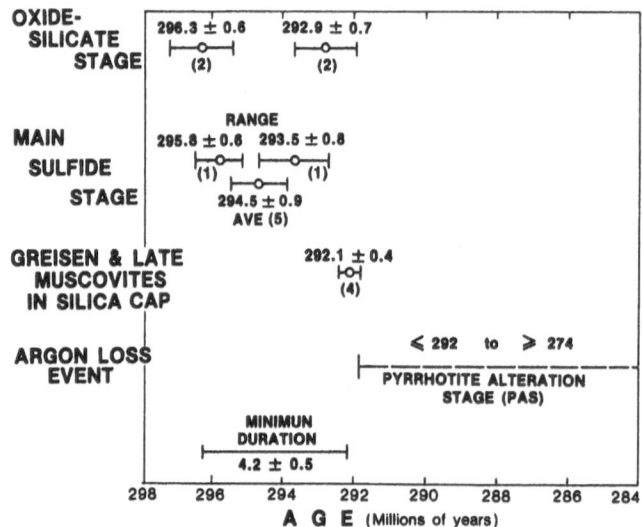

Fig. 83. Detailed $^{40}Ar/^{39}Ar$ geochronology of the Panasqueira tin-tungsten deposit, Portugal (Snee et al. 1988:349). Substantial tin mineralization is associated with the "oxide-silicate stage" (quartz-muscovite-tourmaline-arsenopyrite-cassiterite-wolframite) and the "main sulfide stage"

The generally postmagmatic character of tin mineralization (relative to the immediate granite wall rocks) led Stemprok (1967) to the concept of a para-genetic relationship between tin ore and granitic rocks, which sees the origin of tin-bearing solutions in deep crustal or subcrustal regions. The association of tin deposits and granitic rocks, particularly the frequent localization of tin ore in granite contact zones, was interpreted not as a phenomenon of chemical consanguinity but as a physical consequence of joint lineament-controlled permeability conditions for both granitic melts and hydrothermal solutions. However, this concept of extremely large-scale fluid circulation provides no explanation for the chemical specialization of the ore fluids and of tin granites. In addition, the great increase in solubility of water in granitic melts with increasing pressure probably excludes the coexistence of a melt plus an aqueous fluid phase even at intermediate crustal levels.

6.3 The Magmatic System

A high degree of fractionation is the general feature of granitic rocks associated with tin mineralization. The importance of fractional crystallization is shown by the systematic distribution patterns of major and trace elements. Major elements trend towards the thermal minimum of the Qz-Ab-Or-H_2O system at low pressure, i.e. equilibration to the shallow intrusion environment. Depletion trends of Ca, Mg, Fe, Ti, and complementary depletion and enrichment trends of compatible and incompatible trace elements together with the trend of increasing negative Eu anomaly all indicate fractional crystallization. These chemical evolution trends correlate with temporal and spatial distribution patterns, with the more strongly fractionated melt portions becoming increasingly younger and smaller in volume. Explanations of these phenomena by hypotheses invoking a variable rate of partial melting in the roots of the magmatic system, or by variable melt-restite mixing ratios are inadequate since the linear log-log trace element correlation trends in tin granite suites leave little room for assimilation/contamination as a petrogenetically dominant process. Such processes may play an important role in the deep crust (DePaolo 1981), i.e. in the magmatic evolution at a stage prior to the geochemically documented fractional crystallization sequence.

The synoptic diagrams of Fig. 84 summarize the behaviour of tin in the granite suites discussed earlier. The linear correlation of whole-rock tin content and the two indicators of differentiation, Rb/Sr and TiO_2 in log-log plots is in accordance with the model of a dominant role of fractional crystallization

Fig. 84 (next page). The correlation lines of the granitic sample suites discussed above in the reference system $\log[TiO_2]$-$\log[Sn]$ and $\log[Rb/Sr]$-$\log[Sn]$, respectively. Data on Hercynian tin granites of Portugal from Lehmann (1987). Solid lines are statistically significant at a level of certainty of >99.9 %. Global reference fields from Taylor and McLennan (1985) and Rösler and Lange (1976)

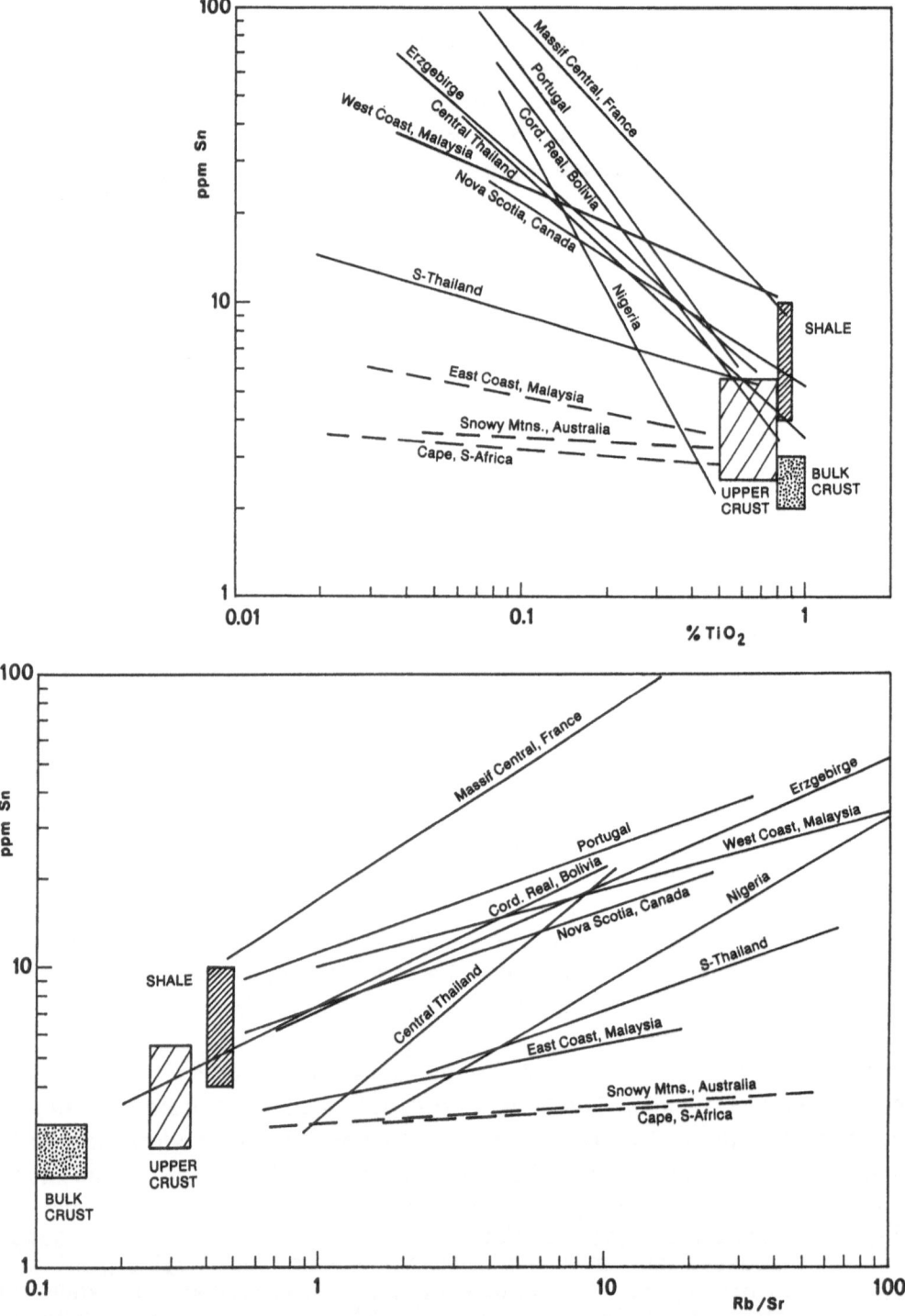

Fig. 84. For legend see previous page

during the magmatic evolution of these rocks (mixing processes result in hyperbolic correlation patterns in log-log space). All fractionation suites can be traced back to Sn levels of ≤10 ppm in their least-evolved portions, most even down to 5 ppm Sn. The assumption that the source material of tin granite suites may already be enriched in tin beyond average crustal levels is therefore not justified. With partial melting rates in anatectic granodiorite-granite systems in the range of 20-50 % (Compston and Chappell 1979) and moderately incompatible behaviour of tin ($\bar{D}_{Sn}$restite/melt = 0.5) bulk-crust material can be expected to yield partial melts with at least 5 ppm Sn.

The comparison with the global reference field for shales suggests correspondingly higher Sn contents in partial melts derived from argillaceous material. The relatively large range of tin contents in shales is, however, in part a consequence of analytical problems with older data derived by optical emission spectrography which tend to be too high (Turekian and Wedepohl 1961; Vinogradov 1962 in Rösler and Lange 1976). Analytical data from neutron activation spectrometry (Hamaguchi et al. 1964), atomic absorption spectrometry (Terashima and Ishihara 1982) and X-ray fluorescence spectrometry (Lehmann et al. 1988) consistently give a shale mean of 3-5 ppm Sn. Partial melting of such material, given an incompatible behaviour of tin, can therefore yield a granitic melt with 5-10 ppm Sn. Pelitic source material is probable for the S-type Hercynian granite suites of western Europe and the Triassic Main-Range granite belt of the SE Asian tin province, which may explain the relatively high tin contents of 5-10 ppm in the least-evolved portions of these granite suites.

The evolution trend of the Cretaceous tin granites of central Nigeria suggests a source material low in tin which appears to correspond to the important mantle component in these rocks (Sn content upper mantle: around 0.8 ppm; Anderson 1983). Efficient titanomagnetite fractionation is, however, feasible as an alternative explanation for the low tin content in early melt portions.

The Rb/Sr-Sn pattern of the central Thailand granites is similar to the Nigerian case (Fig. 84) and results from a mixed sample population which includes some of the chemically primitive Permian volcanic-arc granites of eastern Thailand, which are tin-barren. The extension of this volcanic-arc granite province into peninsular Malaysia (East Coast; "eastern granite belt") is host to some tin deposits in the most evolved granite intrusions. Geochemical and Sr and Nd isotope data indicate involvement of mantle material in this eastern granite belt in the SE Asian tin province (Liew and McCulloch 1985; Cobbing et al. 1986; Darbyshire pers. commun. 1988).

The correlation lines in Fig. 84 have a different slope m which, according to Eq. 12 in Chapter 2.2, is defined by $[\bar{D}_{Sn}-1]/[\bar{D}_{Rb/Sr}-1]$ and $[\bar{D}_{Sn}-1]/[\bar{D}_{TiO2}-1]$, respectively. The slope is therefore related to the magnitude of the bulk tin distribution coefficient $\bar{D}_{Sn}$. Flat correlation lines correspond to $\bar{D}_{Sn} \approx 1$, steep correlation lines point to $\bar{D}_{Sn} < 1$. There is a continuum in slopes for different granitic fractionation suites. The major tin granite suites in Fig. 84 all demonstrate a $\bar{D}_{Sn} < 1$, which explains their geochemical "specialization" in tin. On the other hand, non-tin granites display $\bar{D}_{Sn} \geq 1$. Our examples of non-tin granites are very limited because systematic analytical tin data for such rocks are scarce. The global average for granitic rocks of 3 ppm Sn (Turekian and Wedepohl 1961; Hamaguchi et al. 1964; Grohmann 1965) or 5.5 ppm Sn (Taylor and McLennan 1983) would suggest an average $\bar{D}_{Sn}$ near unity. However, the necessarily low degree of differentiation in large granite systems limits the validity of such an interpretation.

The bulk tin distribution coefficient is essentially controlled by modal composition of the solidifying material in equilibrium with the melt, and possibly also by oxygen fugacity of the system. Fe- and/or Ti-bearing mineral phases are the main tin carriers (Petrova and Legeydo 1965; Kovalenko et al. 1988). Therefore, with Fe-Ti minerals involved in decreasing proportions during progressive fractional crystallization of granitic melts, a dynamic character of $\bar{D}_{Sn}$ may be expected, i.e. a decrease of $\bar{D}_{Sn}$ with growing degree of fractionation. Such a trend was observed in some tin-bearing rocks by Antipin et al. (1981). However, the scatter in the tin data on our sample populations allows no such conclusion. The generally low content of opaque minerals (mainly Ti and Fe oxides and sulphides) in ilmenite-series granitic rocks as opposed to magnetite-series granitic rocks, is shown in Fig. 85, which exhibits a clear compositional distinction between tin-tungsten granites, molybdenum porphyries and copper porphyries.

The petrogenetic classification scheme of magnetite- versus ilmenite-dominated granite suites is geochemically based on different oxygen fugacity fO_2 in these two general melt/rock systems (Ishihara 1981). Some fO_2 data from tin granites and from other ore-bearing granitic environments are depicted in Fig. 86. The chemistry of biotites from the Erzgebirge granite suite allows definition of a trend of decreasing fO_2 towards the latest tin-mineralized granite phases, which spans about two log units from YG 1 to YG 3 and aplite (Förster and Tischendorf 1989). It is interesting to note that the Older Granites of the Erzgebirge (OG 1-3) reveal much higher fO_2 conditions near the HM buffer (Förster and Tischendorf 1989, 1990). These granites are tin-barren but have locally associated molybdenum-tungsten mineralization. This situation is in accordance with the fO_2 data from the molybdenum

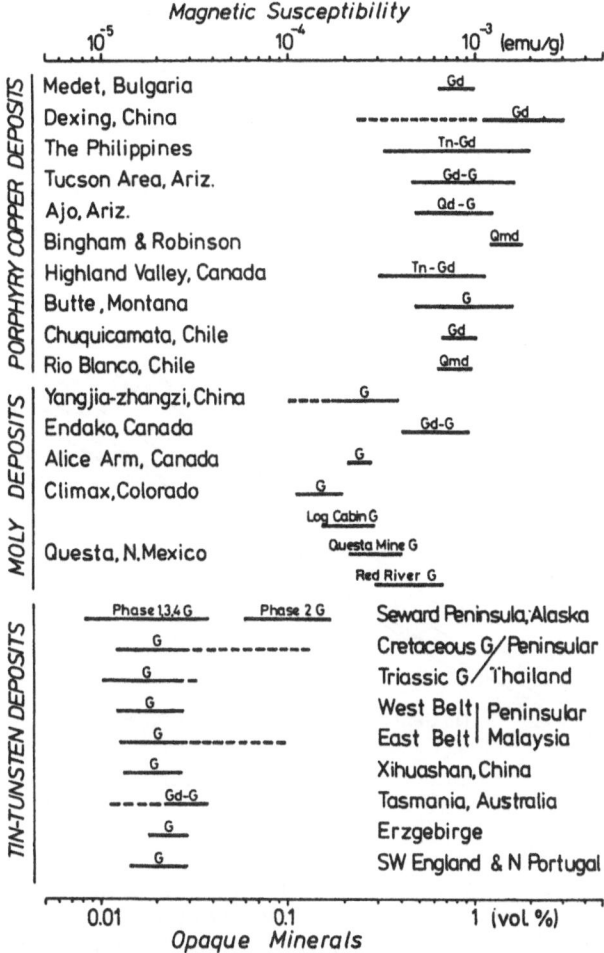

Fig. 85. Modal content of opaque minerals in granitic rocks related to Cu, Mo and Sn-W ore deposits (Ishihara 1981:474). **Tn** tonalite; **Qmd** quartz monzodiorite; **Gd** granodiorite; **G** granite. The limit between ilmenite- and magnetite-series granitic rocks is defined at 0.1 vol% of opaques (Ishihara 1977)

porphyry systems of Questa and Pine Grove, which have redox conditions above the NNO buffer. The fO_2 data from tin ore-related granite phases of the Erzgebirge, from Portugal, Cornwall and from the Seward Peninsula in Alaska plot in between the NNO and QFM buffer lines. The extension of this region towards lower temperature meets the fO_2 conditions typical for the early high-temperature stage in many tin deposits, defined by a position at or below the pyrite-magnetite-pyrrhotite buffer. Oxygen fugacity of tin granite systems at both the magmatic and hydrothermal stage is clearly different from copper porphyry systems (Fig. 86).

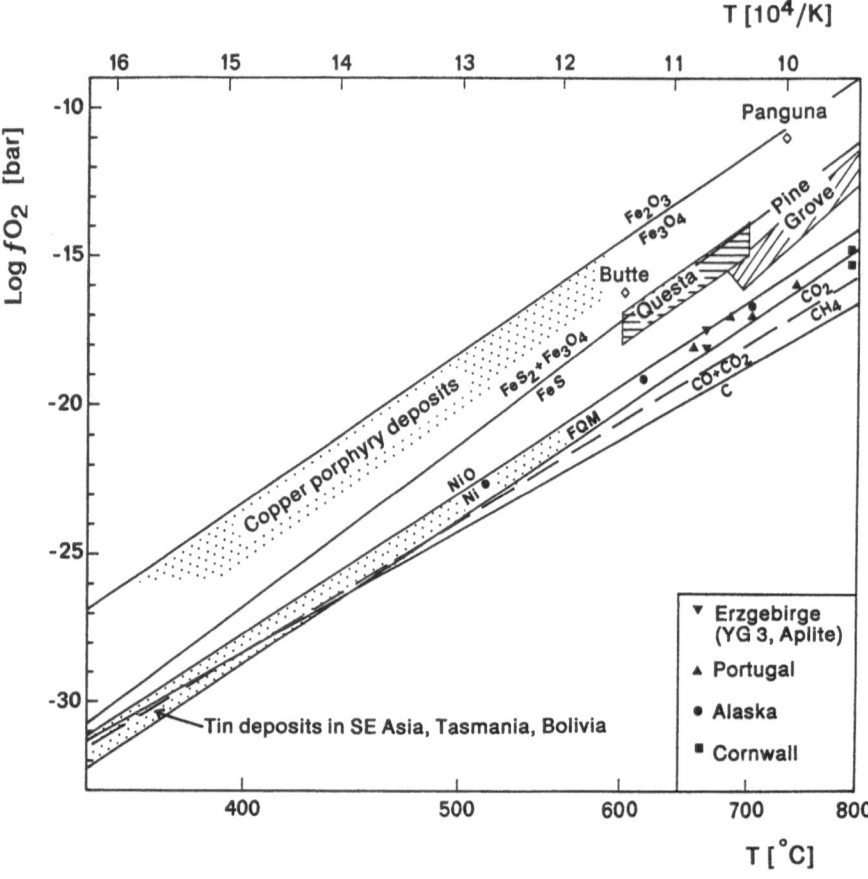

Fig. 86. Temperature and oxygen fugacity conditions of some magmatic-
hydrothermal ore systems as defined by biotite or Fe-Ti-oxide
composition and hydrothermal mineral or fluid inclusion equilibria.
Data for Erzgebirge granites from Förster and Tischendorf (1989),
Portuguese tin granites from Neiva (1976, 1982), Ear Mountain tin
granites, Seward Peninsula, Alaska from Swanson et al. (1988),
Cornubian granites from Charoy (1986) and Stone et al. (1988).
Comparative data for molybdenum porphyries are from the Questa
Caldera, New Mexico (Dillet ánd Czamanske 1987) and the Pine
Grove system, Utah (Keith and Shanks 1988). Copper porphyry data
are from Panguna-Bougainville, Papua New Guinea (Eastoe 1982)
and Butte, Montana (pre-main stage mineralization; McKenzie and
Helgeson 1985). The general compositional field for copper porphyry
mineralization is defined by the equilibrium mineral assemblage
magnetite-hematite (Burnham and Ohmoto 1980). The early
hydrothermal conditions in tin ore formation are defined by a general
position below the pyrite-pyrrhotite-magnetite buffer (cf. Kelly and
Turneaure 1970) and by CO_2/CH_4 ratios in fluid inclusions (Renison
Bell Mine, Tasmania: Patterson et al. 1981; SE-Asian tin belt: Jackson
and Helgeson 1985b; Cornwall: Shepherd et al. 1985). CO_2-CH_4
equilibrium at 1 kbar is from Patterson et al. (1981); other sources for
petrogenetic buffer lines are given in the legend of Fig. 9

Solubility data of tin in silicic melts strongly suggest that crystallization of magmatic cassiterite from a melt containing tin at the ppm level is unlikely (cf. Chap. 2.5). This is particularly unlikely in low fO_2-tin granites, even with tin contents of a highly fractionated melt at the 100 ppm level, but cassiterite saturation may occur in extremely fractionated pegmatite phases. The occasional mention of intramagmatic cassiterite and of intramagmatic tin ore deposits in the literature (Schröcke 1954, 1986) is based on equivocal petrographic evidence, i.e. the idiomorphic habit of cassiterite, which is, however, a general feature of this mineral at any stage of formation and is due to the interfacial energy of cassiterite.

6.4 The Transitional Magmatic-Hydrothermal System

The magmatic and hydrothermal systems in tin granite situations form a continuum characterized by complex transitional phenomena. The transitional stage begins with the differential release of a fluid phase at the solubility boundary of the melt-crystals system as a function of total pressure (level of intrusion), degree of solidification and initial fluid content in the melt. The transitional stage grades below the solidus into the hydrothermal stage. The mobility of the fluid phase depends on the nature and magnitude of permeability, which may lead both to fluid fixation/migration in the intergranular space and to fracture-focussed fluid convection. A consequence of fluid saturation may be fluidization of the residual magma, which may lead to the pervasive emplacement of such material into already crystallized and consolidated granite. Disruptive, secondary magmatic textures of a wide variety may result (see below). A dynamic system composed of multiple intrusions with episodic release of mechanical energy (internal hydraulic fracturing and/or external tectonics) may accumulate fluids in intergranular space which by fracturing may be intermittently tapped and collected on larger structures (reservoir model by Pollard and Taylor 1986). The comagmatic fluid exsolution and accumulation on intergranular space causes chemical and textural convergence between magmatic and hydrothermal mineral formation in granitic rocks, which is probably one of the reasons for the historical discussion of magmatism versus metasomatism in the origin of granites. The petrographical problem of magmatic versus sub-solidus muscovite arises from the same general situation.

The transitional stage in between a magmatic (xtls + melt) and a hydrothermal system (xtls + aqueous fluid), i.e. a system characterized by coexistence of xtls + melt + aqueous fluid (aqueous fluid may under subcritical conditions separate into liquid and vapour phase) will have an abrupt character in the case of synchroneous fracturing and pressure release, and will be longer-lived under quiet pressure conditions. In the first case, favoured by a low-pressure regime at shallow levels, stockwork systems may develop, of which porphyry ore deposits are a fossil relic. The fracture-controlled open system of the porphyry-situation allows rapid and large-scale fluid circulation with concomitant chemical and pysical disequilibria between fluids and wall rocks. In the second case, the fluid phase remains trapped on interstitial spaces or on microcavities. Pegmatitic domains on a micro- to macro-scale may form in which the solidification of highly fractionated melt results in a volumetrically important fluid phase crystallizing in a closed system in equilibrium with the melt phase.

These two end-member scenarios arise as a consequence of the solubility of water in granitic melts (Niggli 1920). The importance of the violent release of a fluid phase in high-level environments was recently emphasized by Whitney (1975), Burnham (1979a,b, 1985) and Burnham and Ohmoto (1980). Supporting textural evidence for fluid exsolution can be found to some degree in any granitic intrusion. Magmatic-hydrothermal phenomena will, however, be most pronounced in strongly fractionated granites where the fluid-accommodating capacity of intergranular spaces may be exceeded. Field evidence for local fluid saturation in tin granites are aplite-pegmatite domains on a cm to m scale in the form of pods, lenses, pipes and veins with repetitive features of textural contrasts (fine-/coarse-grained) and mineral corrosion, and pockets on a microscopic to megascopic scale of quartz-feldspar mosaics. On the microscopic scale, there is a wide variety of textures produced by high-temperature fluid interaction (partially in equilibrium with the magmatic mineral assemblage) such as blebbing of feldspars by albite or secondary quartz, feldspar-quartz myrmekites, micrographic and symplectic intergrowths of quartz, feldspars and muscovite.

The microgranitic fabric of the most evolved tin granite phases is identical to experimental quenching fabrics produced by fluid release (Swanson 1977; Swanson et al. 1988). The fabric in many tin granite intrusions is characterized by two hiatal grain size populations consisting of medium- to coarse-grained quartz, feldspar and biotite porphyroclasts in a fine-grained quartz-feldspar-biotite matrix (microgranitic component). The proportion of megacrystic material to microgranitic groundmass is highly variable and is the reason for the variety of textures in some plutons. There is a continuous spectrum from primary textured "normal" granite to non-megacrystic microgranites, connected by "two-phase variants" (Cobbing et al. 1986). The field aspect of these gradational two-phase variants is characterized in an incipient stage by the thin development of microgranite along grain boundaries in host granite (primary textured granite) and, with an increase in volume of the microgranitic component, leads to disruption of the host granite fabric and corrosion of its mineral constituents by the invading microgranitic matrix. These rocks are usually termed granite porphyries. The alternative term of "megacrystic microgranite" or "two-phase variant" emphasizes the two-phase crystallization process which is interpreted by Cobbing et al. (1986) and Cobbing et al. (in prep.) as a consequence of sudden loss of pressure in a shallow environment causing the residual melt to quench. This results in partial fluidization of the magma and its violent emplacement into, and disruption of the already consolidated primary textured host granite. Fluid penetration beyond the disruption front will produce non-disruptive recrystallization effects in the host granite accompanied by hydrothermal alteration.

The ratio of solid phases (crystals) to liquid (melt + aqueous fluid) will determine the degree of fluidization. According to Wohletz and Sheridan (1979:178), a fluidized system may be defined as "a mixture of particles (solid or liquid) suspended by an upward escaping fluid phase (liquid or gas) so that the frictional force between the fluid and the particles counterbalances the weight of the particles and the whole mass behaves as a fluid". Progressive stages in gas fluidization of solid particles generated by increasing flow rate of the gas phase may change from a fixed bed state, to a quiescent fluidized, to a turbulent fluidized state. The fixed bed state is characterized by an undisturbed grain pattern with fixed contacts between solid particles. Increasing fluid flow leads to a state where individual particles are free to

Fig. 87 (next page). Heterogeneous (secondary) magmatic textures from two-phase granites in the SE Asian tin belt, interpreted as disruptive emplacement of residual melt (fine grained matrix) into partially or wholly crystalline host granite (medium to coarse grained xenocrysts). Photographs from thin sections under oblique polars, length of plates is 20-30 mm. Photographs A and B by courtesy of John Cobbing, Nottingham.
A Xenocrysts of corroded quartzes and feldspars, sieved by quartz blebs, in microgranitic matrix. Dindings Pluton, near Kinta Valley, Malaysia. Sample no. 11 in Cobbing et al. (1986).
B Xenocrysts of corroded quartz, microcline, oligoclase and biotite in microgranitic matrix. Kuala Lumpur Pluton, Malaysia. Sample no. 232/94/416 in Cobbing (1989).
C Megacrystic microgranite composed of xenocrysts of microcline, oligoclase, quartz and brown biotite in fine-grained groundmass of same mineral composition. Tanjungpandan Pluton, Indonesia. Sample no. 217 in Lehmann (1988b)

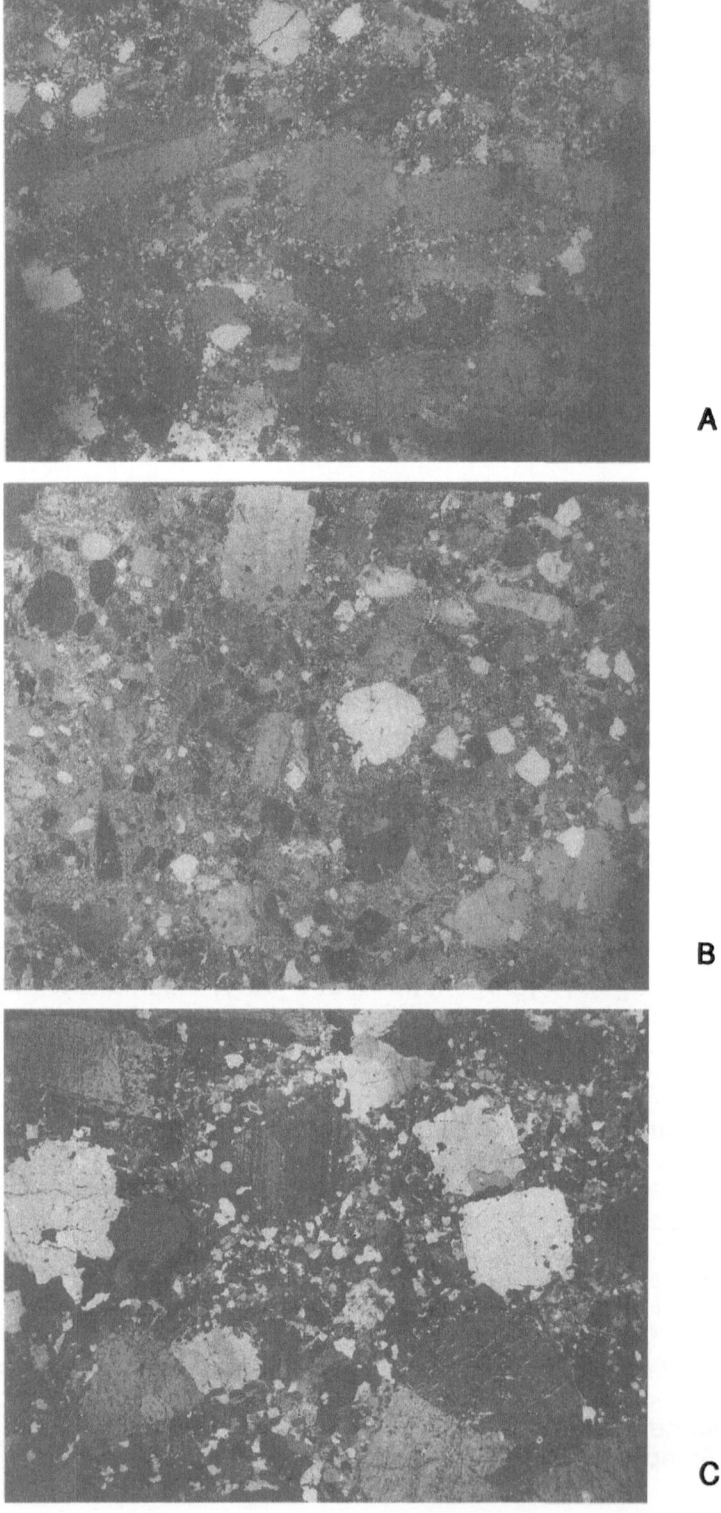

A

B

C

move and where textural integrity is lost. The "fluidizing point" (Leva 1959) is reached when the downward gravity force on the particles is balanced by the upward fluid drag. As fluid flow increases, simple grain separation by increasingly larger spacing is followed by progressively more violent agitation of particles in a turbulent state with abrasion and disaggregation of solid components and the formation of gas bubbles.

The above fluidization phenomena are well known from industrial processes. The most evident application to geological systems is explosive volcanism and breccia pipe formation (Reynolds 1954; Wohletz and Sheridan 1979; McCallum 1985), where turbulent fluidization is an essential feature. The textural patterns of granitic rocks, particularly of highly evolved and boron-rich low-solidus tin granites, suggest a variety of incipient stages of fluidization which are less spectacular and little studied but which contain probably essential physical information on the late magmatic-early hydrothermal evolution of granite plutons.

Secondary magmatic textures are a widespread feature in the granites of the SE Asian tin belt and are commonly associated with tin mineralization (Cobbing et al. 1986; Pitfield et al. 1990). Petrographically, they form a more or less continuous textural suite ranging from slightly disrupted primary texture granites through transitional types to essentially equigranular microgranites. This textural evolution corresponds to a sequence of geochemical evolution (Pitfield et al. 1990). Similar granite textures seem to be important in other tin granite provinces as well and are usually described as granite porphyries (Massif Central: Aubert 1969; Cornwall: Hall 1974; Bolivia: Lehmann 1979). Examples for secondary magmatic textures in tin-bearing granites in the SE Asian tin belt are given in Fig. 87.

Fluid release from within the magma as a result of second boiling provides an explanation for the fracture pattern in high-level granitic intrusions which is centred on late and most evolved intrusion phases; a situation typical of copper, as well as tin-tungsten and molybdenum porphyries/granites. The high salinity and stable isotope composition of early hydrothermal fluids in these deposits are in favour of a magmatic origin of the ore fluids which during the cooling history of the magmatic-hydrothermal systems become dominated by external fluids.

The transitional magmatic-hydrothermal stage (i.e. coexistence of crystals, melt and aqueous fluid) is the key condition for most models of pegmatite formation (Smith 1948; Jahns and Burnham 1969; Cerny 1982). Extreme variations in grain size in miarolitic rare-metal bearing pegmatites, particularly

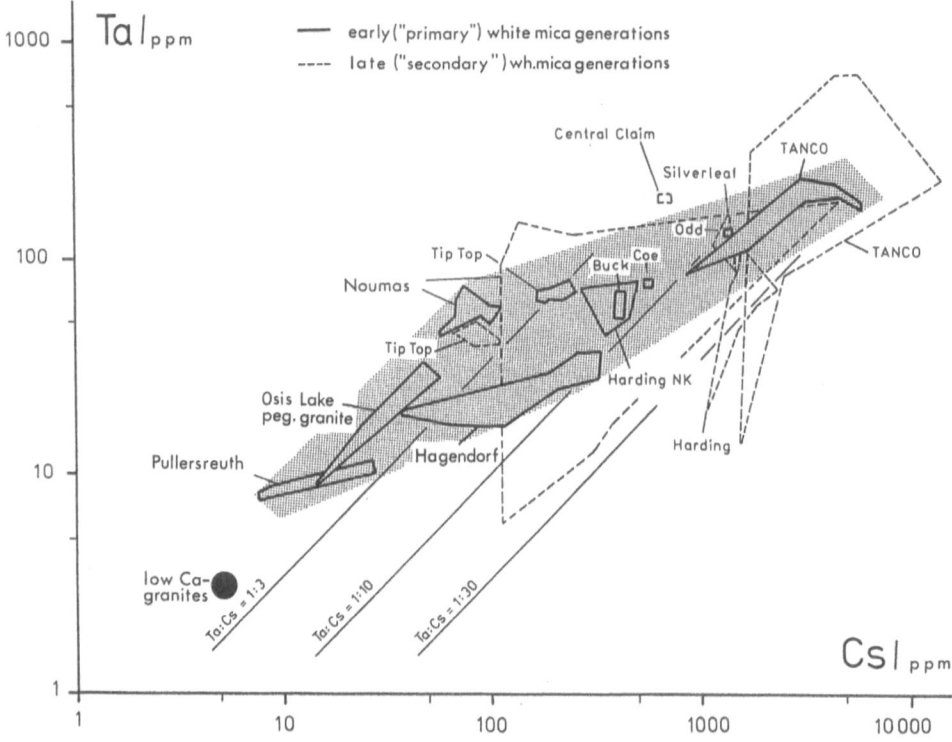

Fig. 88. Ta-Cs variation diagram for white micas from various granitic pegmatites (Gaupp et al. 1984; redrawn in Möller 1989:110). Shaded compositional field delineates data on early (primary) micas. Solid circle locates whole-rock composition of low-Ca granites. Strong Ta mineralization is associated with the Tanco, Silverleaf, Odd West, Buck Claim, Coe Claim, Central Claim pegmatites (Winnipeg River pegmatite district, Manitoba, Canada); moderate Ta mineralization is associated with the Harding pegmatite (New Mexico, USA), Tip Top pegmatite (Black Hills, South Dakota, USA), and Noumas pegmatite (Namaqualand, RSA); the ceramic pegmatites of Hagendorf and Pullersreuth (Oberpfalz, Bavaria, Germany) have no metal mineralization

the typical association of aplitic and very-coarse grained domains, point to crystallization from two physically different fluids in the sense of the classical model by Jahns and Burnham (1969). The clear distinction between a granitic silicate liquid (melt) and a coexisting, comparatively low-density, saline aqueous fluid must, however, not be the case in boron-rich granite pegmatite systems, in which the pegmatitic fluid appears to have evolved continuously from a magmatic towards a hydrothermal state (London 1986).

A high degree of fractionation in Sn- and Ta-bearing pegmatites is indicated by the same systematic enrichment and depletion patterns which are typical of tin granites. Compositional fields of white micas from a variety of pegmatites are given in Fig. 88 in terms of Ta versus Cs. Corresponding whole-rock data from alkalifeldspar aplogranite stocks of the Erzgebirge (Tischendorf 1989) would plot in an intermediate position in the compositional field for pegmatitic micas in Fig. 88.

Ratios of chemically coherent elements such as K/Rb, Ga/Al, Ta/Nb and Zr/Hf reach anomalous values in rare-metal pegmatites and follow the same trend line as highly fractionated granites (Cerny, 1982, 1989; Möller 1989). Cassiterite and zircon samples from pegmatites are characterized by Zr/Hf ratios of 4 ±2, those from hydrothermal vein deposits have ratios of 28 ±10, whereas average zircons and rocks of terrestrial and meteoritic origin have a Zr/Hf ratio of 43 ±8 (Möller and Dulski 1983; Cerny et al. 1985). The extremely low Zr/Hf ratio in cassiterite of pegmatitic origin points to a degree of fractionation F of less than 10^{-4} (Möller and Dulski 1983).

The extremely fractionated nature of Sn-Ta enriched pegmatite melts/fluids (enriched in other incompatible components such as H_2O, F, B, Li, Rb, Cs, Be) results in low viscosity and density of such systems which, given suitable

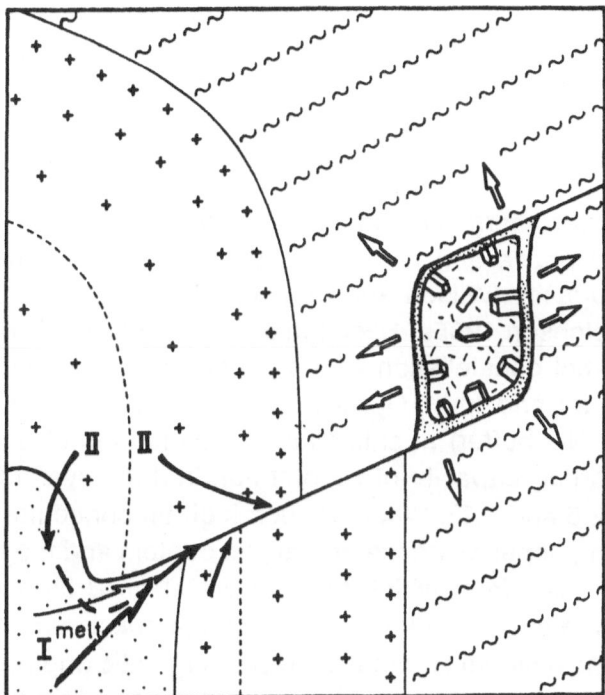

Fig. 89. Schematic two-component model of granitic pegmatite formation (Möller 1989:120). Fluid I is a late residual hydrous granitic melt, fluid II represents the intergranular liquid from crystallized granite which mixes continuously with fluid I

channelways, may migrate upwards and outwards of their granitic parent intrusion along fractures or zones of weakness. The most fractionated melt batches are the most mobile, which is seen as the reason for the zonal distribution of pegmatite types and associated metal mineralization in the roof of granite intrusions (Cerny 1989). The contact between different granite units or granite and country rock can be a favorable zone for pegmatitic fluid migration towards the most apical parts of granite intrusions. In the Erzgebirge, pegmatite zones along such contacts are known under the traditional term of "stockscheider", which consists mainly of coarse-grained K-feldspar, quartz, Li-bearing mica, and - locally - columnar aggregations of topaz (pycnite).

Alteration haloes around pegmatite bodies, heterogeneous $^{18}O/^{16}O$, D/H, $^{87}Sr/^{86}Sr$ and disturbed Rb-Sr patterns, and fluid inclusion evidence (Roedder 1984) point to a partly open system during pegmatite crystallization. The model of pegmatite formation by Möller (1989) envisions a combination of both a process of internal magmatic evolution (granitic and pegmatitic fluids) through fractional crystallization and a second process of mixing with intergranular fluids from the crystallized granite parent (Fig. 89).

6.5 The Hydrothermal System

The mobility of tin in hydrothermal solutions is, of course, independent of the origin of the aqueous phase and is essentially controlled by temperature, salinity and oxygen fugacity (Wilson and Eugster 1984; Eugster and Wilson 1985; Jackson and Helgeson 1985a; Haselton and d'Angelo 1986; Kovalenko et al. 1986). The optimal constellation of these parameters is given at high salinity and temperature and low fO_2. An aqueous phase in a tin granite system (at or below NNO buffer) at subsolidus temperatures of 500-600 $^{\circ}$C will have a tin-transporting capacity of several hundred to several thousand ppm Sn (cf. Chapts. 2.6 and 2.7). Whether such high tin concentrations can be reached at any point, depends on the primary tin content and the fixation of tin in the percolated rocks, from which tin may be released by a variety of congruent or incongruent dissolution and exchange reactions. It can be expected that part of the residual aqueous fluids of magmatic origin is trapped on grain boundaries or microcavities in the rock, thus providing a vast fluid reservoir (Pollard and Taylor 1986). These interstitial fluids can migrate to higher structural levels under the influence of thermal and pressure gradients, or can be flushed out by external fluids such as ingressing meteoric water.

Leaching of biotite together with muscovitization and chloritization may be another mechanism of tin release, suggested early on by Barsukov (1957).

Repeated dissolution of hydrothermal cassiterite and crystallization of high-grade tin ore is favoured by a fluid system which is not entirely buffered by its granitic wall rocks (equilibrium assemblage quartz-feldspar-muscovite) and where a decreasing pH during fluid cooling allows high tin contents in the hydrothermal fluid down to temperatures of 200-250 $^{\circ}$C (Heinrich and Jaireth 1989). Low-pH conditions of a fluid-buffered hydrothermal evolution are possible in spatially restricted circulation systems such as quartz veins or greisen environments in which feldspar was previously destroyed.

Efficient hydrothermal leaching, transport and deposition of tin apparently need a magmatically tin-preenriched reservoir on which the hydrothermal extraction process must be focussed. This is in contrast to the metallogeny of gold and base metals, for instance, and must be the reason why economic tin concentrations in high-temperature, low-pH and low-fO_2 hydrothermal systems not related to a tin-enriched rock volume are so rare. Favourable conditions for tin leaching and transport can, for example, be expected in such parts of hydrothermal systems at mid-ocean ridges where the fluid is rock-buffered by mantle material at a redox state up to four log units below FMQ, i.e. at the iron-wüstite buffer (Bryndzia et al. 1989). Although cassiterite occurrences from such a setting have been reported, these are very small tin concentrations only, compared to the much more efficient extraction and enrichment processes for base metals (Dmitriev et al. 1971; Jankovic 1972). The locally high tin content in volcanogenic massive sulphide deposits, which permits in some cases a by-production of tin, is probably understandable as the result of such a low-fO_2 environment. A countercurrent distribution pattern of tin and gold can be predicted as a consequence of the complementary solubility behaviour of the two redox couples Au^+-Au^0 and Sn^{2+}-Sn^{4+}, i.e. tin fixation (cassiterite precipitation) under relatively more oxidized conditions and gold precipitation in more reduced parts of the same hydrothermal system (given a sulfidation state inside H_2S stability field).

The fact that tin deposits have such a strong preference for highly fractionated granitic rocks, in which tin is enriched about one order of magnitude over average crust, is not fully explainable by the tin specialization alone of their host environments. Otherwise, small-scale tin deposits should be much more abundant in non-granitic terranes. Highly fractionated granitic systems apparently not only provide the chemical inventory for tin ore formation, but also provide this inventory in a framework which allows efficient extraction and hydrothermal redistribution. This situation is probably a consequence of the interstitial fluid release during granitic crystallization.

The formation temperature of cassiterite in tin ore deposits is commonly in the range of 300-500 $^\circ$C as noted in the first summary study by Little (1960) and later confirmed by numerous investigations. Occurrences of colloidal wood tin (hydro-cassiterite) in Mexico seem to have formed at a temperature as low as about 150 $^\circ$C (Pan 1974). Salinity data from fluid inclusions in cassiterite and associated quartz vary widely between 1 and >50 wt% NaCl-equivalent, with a typical average range of 5-20 wt% (1-4 m NaCl) (Bolivia: Kelly and Turneaure 1970; Grant et al. 1977; Erzgebirge: Durisova et al. 1979; Thomas 1982; Tasmania: Patterson et al. 1981; Portugal: Kelly and Rye 1979; Cornwall: Jackson et al. 1982; Nigeria: Kinnaird 1985; and general review in Roedder 1984). The hydrothermal conditions during cassiterite crystallization, as derived from the study of fluid inclusions, correspond to the solubility range of a few to a few hundred ppm Sn.

It should be borne in mind that fluid inclusion data record the depositional regime of the minerals investigated, i.e. the final stage in the evolution of an ore fluid. The stable isotope data from tin ore deposits uniformly indicate that the tin-mineralizing fluids were equilibrated isotopically (and probably chemically) with a hot (T > 400-500 $^\circ$C) granitic mineral assemblage, prior to their transport into a cooler ore deposition site (Kelly and Rye 1979; Grant et al. 1980; Patterson et al. 1981; Pollard and Taylor 1986; Sun and Eadington 1987; Thorn 1988).

Oxygen fugacity during tin ore formation can be reconstructed from the mineral paragenesis and composition of fluid inclusions. A worldwide feature of tin deposits is an early high-temperature stage of pyrrhotite ± pyrite followed by a late stage (post-cassiterite mineralization) of pyrrhotite alteration in which pyrrhotite is pseudomorphically transformed into pyrite/marcasite ± siderite (see Kelly and Turneaure 1970). Pyrite blastesis tends to completely extinguish the primary pyrrhotite fabric. The occurrence of magnetite (often as inclusions in cassiterite and in fluid inclusions) together with pyrrhotite ± pyrite defines an upper fO_2 limit at the pyrrhotite-pyrite-magnetite buffer. The lower fO_2 limit is given by the ratio CO_2/CH_4 in fluid inclusions. This ratio is >1 for most tin deposits (temperature: 350-400 $^\circ$C). In many cases CH_4 is not detectible, in some deposits CO_2/CH_4 is around unity (Tikus, Indonesia: Schwartz and Surjono 1988; Renison Bell, Australia: Patterson et al. 1981; Sa Moeng, Thailand: Khositanont in prep.). These observations suggest an fO_2 interval for cassiterite formation in between the pyrrhotite-pyrite-magnetite (at 350 $^\circ$C nearly identical to NNO) and the QFM

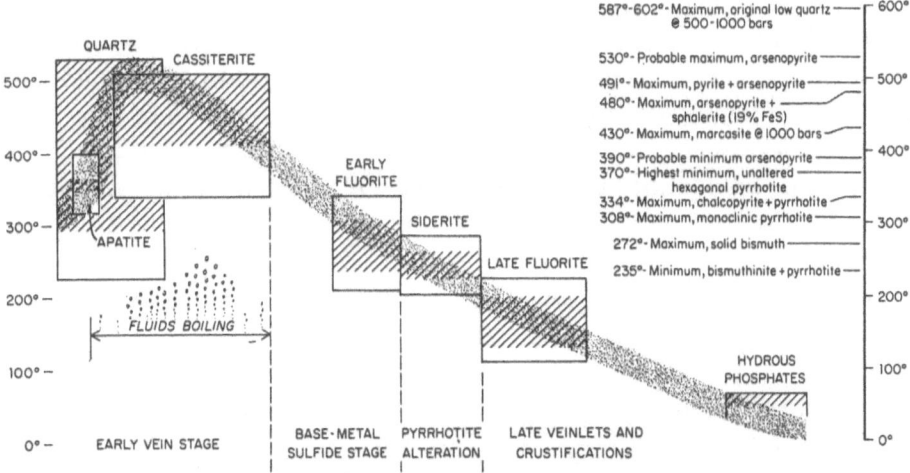

Fig. 90. Generalized temperature evolution of tin-tungsten deposits in the Bolivian tin belt, derived from fluid inclusion data and from mineral equilibria (Kelly and Turneaure 1970:673). Typical mineral assemblages are (Kelly and Tuneaure 1970; Hanus 1982; Wolf and Sanchez 1976). 1 Early vein stage: quartz-muscovite-tourmaline-apatite-cassiterite-wolframite-löllingite/arsenopyrite-fluorite-siderite-chlorite. 2 Base-metal sulphide stage: pyrrhotite-arsenopyrite-sphalerite-bismuth-bismuthinite-chalcopyrite-stannite-galena.
3 Epithermal stage: pyrite-marcasite-siderite (reaction paragenesis from pyrrhotite alteration); complex sulphides-stibnite, Bi-Co-Ni-Ag-U sequence and gold, fluorite-baryte-phosphates-limonite-gold-silver (low-temperature remobilization)

reference buffers. This is the same interval which, at higher temperature, characterizes the crystallization of tin granites (Fig. 86), and is probably a consequence of the broad-scale control of the hydrothermal system by its granitic wall rocks, at least in an early stage with a low water/rock ratio . A trend of increasingly external influence on the hydrothermal system can move the fluids and their wall rocks at a later stage into the hematite stability field.

Figure 90 summarizes the typical temperature evolution and mineralogical sequence in hydrothermal tin deposits. The diagram is from the classical work by Kelly and Turneaure (1970) who studied over fifty tin and tungsten deposits in Bolivia which in spite of different ages and genetic types gave a uniform evolutionary pattern. This pattern, although in individual deposits commonly not developed in full, provides a general framework with respect to temperature and mineral association in tin ore formation.

Fluid oversaturation and precipitation of cassiterite in those paleo-solution channels which are today's tin deposits is a consequence of one or several processes which, according to Figs. 13 and 14, lead to cassiterite formation: decrease in temperature and/or chloride activity, increase in fO_2 and/or pH. All these processes can be caused and influenced by mixing with meteoric water and interaction with wall rocks. A predominantly meteoric water component is typical for the later stages of hydrothermal evolution in tin ore systems, as seen in D/H isotope patterns in many localities (Cornwall: Jackson et al. 1982; Portugal: Kelly and Rye 1979; Thailand: Lehmann 1988a; Bolivia: Grant et al. 1980; etc.).

Cassiterite precipitation according to the general equilibrium

$$Sn^{2+} + H_2O + \tfrac{1}{2}O_2 = SnO_2 + 2H^+ \qquad (26)$$

is redox- and pH-dependent and must therefore be accompanied by H^+- and H_2-consuming processes. Acid neutralization is most effectively achieved in carbonate rocks during the formation of skarn or sulphide replacement tin deposits by the model reaction

$$Sn^{2+} + CaCO_3 + H_2O = SnO_2 + Ca^{2+} + CO_2 + H_2. \qquad (27)$$

Feldspar-destructive wall rock alteration seems to be the dominant H^+-consuming reaction in non-carbonate rocks in which muscovitization is an important feature, such as:

$$Sn^{2+} + 3(Na,K)AlSi_3O_8 + 2H_2O =$$

$$SnO_2 + KAl_3Si_3O_{10}(OH)_2 + 6SiO_2 + 2Na^+ H_2. \qquad (28)$$

The commonly observed paragenetic mineral assemblage cassiterite + arsenopyrite has been proposed by Heinrich and Eadington (1986) as a wall rock-independent redox buffer. Precipitation of arsenopyrite by reduction of As^{3+} complexes may couple with oxidation of Sn^{2+} complexes to precipitate cassiterite:

$$3Sn^{2+} + 2H_3AsO_3 + 2Fe^{2+} + 2H_2S = 3SnO_2 + 2FeAsS + 10H^+ \qquad (29)$$

An alternative redox couple for fluid-buffered cassiterite precipitation in CO_2 rich fluids is possible through the equilibrium $CO_2 + 4H_2 = CH_4 + 2H_2O$ (Giggenbach 1980).

6.6 Synthesis

Tin ore deposits are part of fossil hydrothermal systems centred on highly fractionated late phases of extended granitic fractionation suites. These hydrothermal systems are developed in apical high-level portions of granitic intrusion series, and their association with granitic magmatism typically developed in active uplift regions (Pitcher 1979, 1987), makes them extremely sensitive to erosional processes. Primary tin ore deposits, in association with granitic rocks of collisional belts and in back-arc regions, are therefore mostly Mesozoic-Cenozoic in age, whereas tin granite systems in anorogenic intracratonic settings can be traced back to Archaean times. A similar situation can be found for copper and molybdenum ore deposits in association with granitic magmatism.

The granitic phases/subintrusions in spatial, temporal, and chemical relationship to tin ore deposits (i.e. tin granites) are highly fractionated. According to systematic trace element distribution patterns in these subunits and in associated larger granite systems, fractional crystallization seems to be the dominant petrogenetic process controlling magmatic evolution. Convective fractionation, i.e. convection of fluid away from crystals, provides an effective crystal-melt separation mechanism (Rice 1981; Sparks et al. 1984). The multiphase nature of the intrusion systems and their systematic enrichment patterns in incompatible trace elements towards the youngest and volumetrically smallest phases point to a crystallization process from the plutonic margins inwards (Groves and McCarthy 1978) and episodic tapping of an increasingly fractionated and buoyant magma chamber below the gradually deepening solidification front.

Sr and Nd isotope data indicate for most tin granite suites an origin predominantly by partial melting of crustal material; in some tin provinces a substantial mantle component seems to be involved. The starting material of the granitic fractionation suites has no anomalous contents in tin as compared to average pelitic material. The tin specialization of tin granites is a consequence of their magmatic evolution by fractional crystallization, and crustal thickness may be an important parameter in influencing the magma residence time and its spatial evolution. This is suggested by the more continental setting of tin granite systems as compared to the chemically less evolved copper porphyries. The origin from mantle or crustal sources is probably secondary to the importance of fractionation mechanisms because the ratio of tin content of bulk crust to upper mantle is only two to four.

Oxidation state (conventionally described by oxygen fugacity fO_2) seems to control the bulk tin distribution coefficient $\bar{D}_{Sn}$(crystals/melt), i.e. degree of magmatic tin enrichment during fractional crystallization, in influencing the Sn^{4+}/Sn^{2+} ratio in the melt. Contrary to the behaviour of molybdenum which is enriched in molybdenum porphyries at $fO_2 > NNO$ (Ni-NiO buffer), tin granites are characterized by conditions of $fO_2 < NNO$. This difference is reflected by the magnetite-series mineralogy in granitic systems associated with molybdenum ore deposits, whereas tin-bearing systems are associated with ilmenite-series granitic rocks (Ishihara 1981). Other factors influence $\bar{D}_{Sn}$ as well, which are again dependent on degree of fractionation. As crystallization proceeds, $\bar{D}_{Sn}$ can be expected to decrease towards late crystallization stages, due to a modal increase in mineral phases with low partition coefficients D_{Sn}. The increasingly depolymerized melt structure during fractional crystallization as a result of the enrichment in H_2O, B, F may, in addition, further decrease $\bar{D}_{Sn}$ by reducing individual mineral partition coefficients D_{Sn} (Antipin et al. 1981).

Degree of fractionation and oxidation state are the two parameters of prime importance for magmatic tin enrichment in granite suites, and have a regional control over the formation of tin provinces. Degree of partial melting seems to be of lesser importance in granite systems because the physical process of silicic melt separation during anatexis probably limits the relevant range of degree of partial melting to 20-50 % (Compston and Chappell 1979; Thompson and Connolly 1990). Low oxygen fugacity in granitic melt could be provided by source material with high Fe^{2+}/Fe^{3+} ratio and/or high carbon or S^{2-} content. Pelitic sedimentary sequences on the order of 10 km of stratigraphic thickness are typical for the basement of many tin provinces (SE Asia, Bolivia, Cornwall, Portugal). Pelitic source rock material makes little difference to the magmatic tin distribution patterns in tin granites as compared to other source rocks. However, the extreme boron enrichment of most tin granite systems may reflect the distinct boron enrichment of average shale (100-200 ppm B) as compared to bulk crust (10 ppm B), whereas the corresponding tin data are 2.5 ppm (bulk crust) and 3-5 ppm (shale) (Taylor and McLennan 1985). An indication for pelitic source rock material in most (but not all) tin granite suites is the generally peraluminous S-type character of these granites, which contrasts with I-type fractionation suites associated with F-dominated molybdenum ore systems.

The hydrothermal tin ore system is a continuation and result of the magmatic evolution trend, with shallow intrusion level providing high permeability and fluid circulation. The exsolution of a chloride-bearing fluid phase from a crystallizing granitic melt is the necessary consequence of anhydrous crystallization of a hydrous melt (Holland 1972). The exsolved fluid phase can be accommodated and stored by the intergranular space in little fractionated magma portions. In higher fractionated melt portions, larger physical domains of a fluid phase will form during solidification, accompanied by focussed development of mechanical energy during retrograde boiling (Burnham 1985). The symmetry of structurally controlled permeability patterns centred on apical portions of highly evolved granite stocks argues for such a situation which is particularly developed in subvolcanic settings (porphyry-type stockworks, pervasive brecciation, breccia pipes).

A continuum may be expected in between the more static fluid storage in intergranular spaces and explosive fluid-melt phenomena. A two-phase crystallization pattern is typical of highly fractionated granites, with a primary medium- or coarse-grained granitic fabric forcefully invaded and partly corroded by fine-grained material of similar composition. The ratio between both phases is highly variable on a metre- to centimetre-scale. The bimodal grain distribution and locally pervasive disruption of the primary fabric by very mobile late-magmatic, fine-grained material with development of blastomylonitic textures suggests coexistence of crystals and melt with a fluid phase under $P_{fluid} > P_{lithostatic}$. The identity of the resulting rock as a porphyroclastic or megacrystic microgranite and its wide distribution in granite provinces has only recently been demonstrated by regional granite mapping programmes in SE Asia (Cobbing et al. 1986). Two-phase rocks under the conventional description as granite porphyry or porphyritic granite are generally well known to be associated with tin mineralization.

The mobilization of the magmatically developed tin potential in a highly fractionated granite phase is possible both in the earliest stage of magmatic fluid release as well as during later invasion of the cooling intrusive system by meteoric water. Initially, there will be a separation of both fluid systems, due to the extremely low permeability of a melt. Fracturing during the sub-solidus evolution will lead to superposition and mixing of both fluid systems. The time span of tin mineralization in some ore deposits is on the order of several million years, which points to the existence of long-lived hydrothermal fluid circulation at high temperatures ($>300\,^{\circ}C$).

The hydrothermal mobility of tin is physically controlled by the nature and extent of permeability and available fluid volumina during the subsolidus cooling history of a tin granite. Both factors are favoured in high-level environments. The chemical parameters pH, fO_2 and chloride activity control the tin carrying capability of a solution. Fluid-buffered low-pH conditions in hydrothermally altered rock portions (feldspar unstable) can sustain high tin contents in hydrothermal solutions even at 200 $^{\circ}$C (Heinrich and Jaireth 1989).

Size and primary tin content of a granite system define the maximum quantity of extractable metal. The greatest hydrothermal tin enrichments in individual ore systems are on the order of 10^6 t Sn (ore deposit plus primary dispersion halo; e.g. Llallagua, Bolivia, or Altenberg, Germany). Tin granite subintrusions reach average primary tin levels of around 30 ppm. The hydrothermal depletion of a tin-enriched magmatic system must therefore comprise a large rock volume of up to 40 km^3, when an extraction rate of one third (10 ppm Sn) is assumed. The whole magmatic fractionation system must be at least ten times larger in order to produce a highly fractionated tin granite subintrusion. The limiting assumption of Rayleigh fractionation with $\bar{D}_{Sn}=0$ gives a minimum size of the total magmatic system of 400 km^3. More realistic conditions of a less perfect fractionation mechanism require a total melt volume on the order of 1000-2000 km^3. This is the size of magma chambers known to be associated with some caldera eruptions (Hildreth 1981). The granite batholiths underlying the Erzgebirge and Cornwall tin provinces are about 50 times larger (Tischendorf 1989; Willis-Richards and Jackson 1989).

The process of hydrothermal tin extraction results in open-system distortions of primary magmatic tin distribution patterns (scatter distributions with tin deficiencies). Often however, hydrothermal tin depletion cannot be identified clearly, which suggests relatively small extraction rates not distinguishable from the internal scatter in (pseudo)-magmatic tin enrichment trends.

A synoptic view of the magmatic and hydrothermal evolution of a tin-bearing granite system is given in Fig. 91 in conceptual analogy to copper porphyry systems (Burnham 1979a):

1. Intrusion of a granite body: sidewall crystallization combined with a convective crystal-melt separation mechanism results in progressive fractionation of residual liquid. Thermal anomaly induces external meteoric-hydrothermal convection system. Solidifying magma portions release fluid into intergranular space.

2. Subintrusions of higher fractionated and less dense magma blobs develop in response to buoyancy and, possibly, tectonic squeezing. Solidus temperature decreases with increasing degree of fractionation, i.e. with enrichment in H_2O, F, B. Retrograde boiling during crystallization leads to magmatic fluid release with concomitant release of mechanical energy (stockwork and breccia formation); convergence between magmatic-

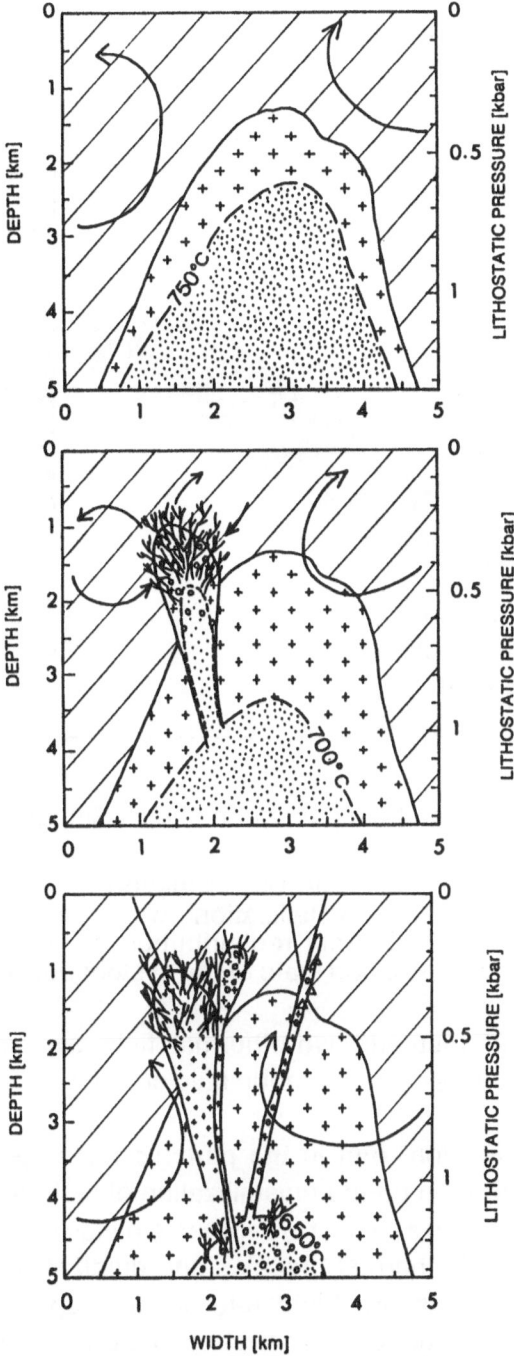

Fïg. 91. Three stages in the schematic evolution of a tin granite system (5 x 5 km sections) in the upper part of a larger granite batholith. Arrows indicate trends of large-scale fluid movement. Broken-line contour in granite pluton marks solidus zone. Circle pattern represents zone of vapour release. For further explanation see text

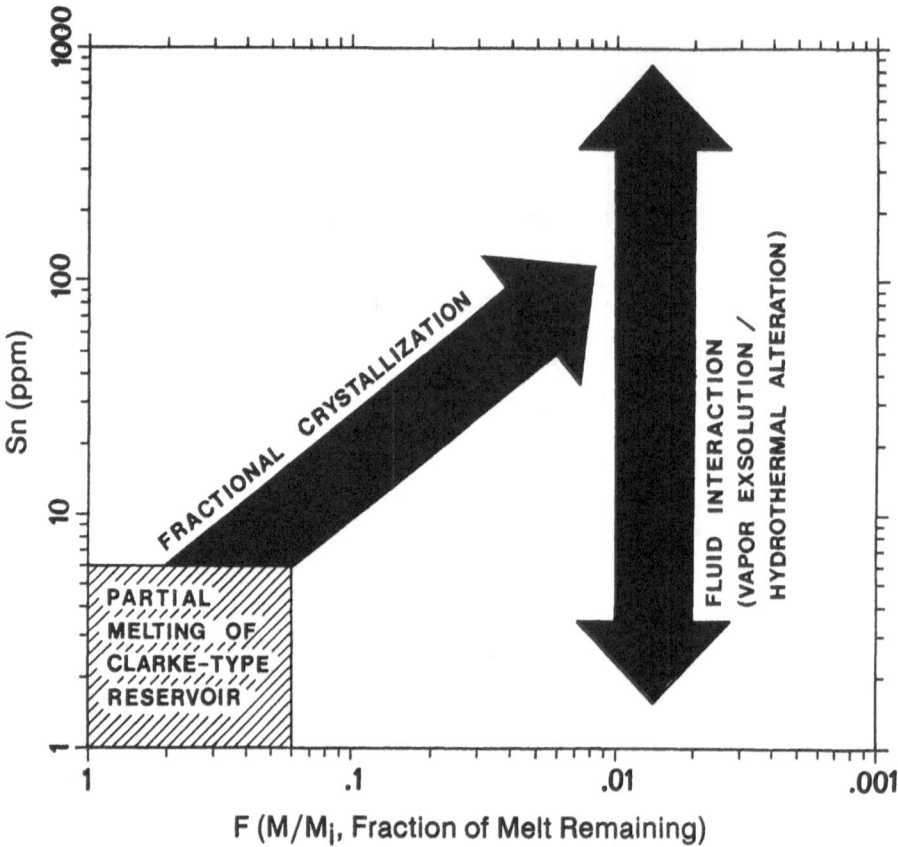

Fig. 92. Metallogenic model of tin ore formation: magmatic tin enrichment through fractional crystallization with $\bar{D}_{Sn}$(xtls/melt) <1, and subsequent hydrothermal redistribution of tin. Both processes are favoured by an oxidation state of the system below the NNO buffer

hydrothermal and meteoric-hydrothermal fluid systems with temporally increasing dominance of the meteoric fluid component.

3. Retreating solidification front in the plutonic main system with tendency towards fluid saturation during crystallization of residual melt; subintrusion of microgranitic stocks and aplite/pegmatite dykes with individual fluid and energy release: hydrothermal breccias and fluidization phenomena in the crystal-melt-vapour system. Fluid release at deeper levels and thermal contraction produce deeply penetrating tectonic structures. Increasing ingress of meteoric fluids and flushing out of interstitially stored residual magmatic fluids during decreasing temperature results in greisenization, potassic, sericitic, argillic, propylitic alteration etc. (systematic alteration patterns in analogy to other granite-related hydrothermal systems).

Focussing of the hydrothermal system according to local permeability conditions and cassiterite deposition as consequence of local physical and chemical gradients leads to the variety of the tin ore deposit spectrum (pegmatite, greisen, stockwork/vein, skarn/sulphide replacement).

The same general model is condensed in Fig. 92 under a geochemical point of view. The metallogenic model has two basic features (Lehmann and Mahawat 1989):

1. Magmatic tin enrichment trend: fractional crystallization of granitic melt with a bulk tin distribution coefficient $\bar{D}_{Sn}$(xtls/melt) <1. This condition is realized in peraluminous, ilmenite-series granitic rocks with an oxygen fugacity below NNO, which in turn is a precondition for effective hydrothermal depletion of magmatically tin-enriched rock volumes.

2. Hydrothermal tin redistribution pattern: magmatic fluid release during solidification of the highly fractionated tin granite system combined with interaction of the solidified granite with meteoric-hydrothermal fluids. Loading of the hydrothermal system with magmatically enriched tin, and cassiterite deposition according to local physicochemical gradients.

References

Abbey S (1983) Studies in "standard samples" of silicate rocks and minerals 1969-1982. Pap Geol Surv Canada 83-15: 1-114

Adam JWH (1960) On the geology of the primary tin-ore deposits in the sedimentary formation of Billiton. Geol Mijnbouw 39: 405-426

Agricola G (1546,1956) De ortu et causis subterraneorum libri V. Basel (Froben), 81 pp (Die Entstehung der Stoffe im Erdinnern) In: Georgius Agricola - Ausgewählte Werke, 3. VEB Dtsch Verlag Wiss, Berlin, pp 83-187

Ahlfeld F (1931) Über Tracht und Genesis des Zinnsteins. Fortschr Miner Krist Petr 16: 47-49

Ahlfeld F (1958) Zinn und Wolfram. Enke, Stuttgart, 210 pp

Ahlfeld F, Schneider-Scherbina A (1964) Los yacimientos minerales y de hidrocarburos de Bolivia. Bol Dept Nac Geol < La Paz> 5: 1-388

Albuquerque CAR de (1971) Petrochemistry of a series of granitic rocks from northern Portugal. Bull Geol Soc Amer 82: 2783-2798

Alderton DHM, Pearce JA, Potts PJ (1980) Rare earth element mobility during granite alteration: evidence from southwest England. Earth Planet Sci Lett 49: 149-165

Aleva GJJ (1985) Indonesian fluvial cassiterite placers and their genetic environment. J Geol Soc London 142, 815-836

Amstutz GC (1959) Syngenetic zoning in ore deposits. Proc Geol Assoc Canada 11: 95-113

Anders E, Grevesse N (1989) Abundances of the elements: meteoritic and solar. Geochim Cosmochim Acta 53: 197-214

Anderson DL (1983) Chemical composition of the mantle. Proc 14th Lunar Planet Sci Conf, part 1, J Geophys Res 88 (suppl): B41-B52

Antipin VS, Kovalenko VI, Kuznetsova AI, Persikova LA (1981) Distribution coefficients for tin and tungsten in ore-bearing acid igneous rocks. Geochem Intern 18, 1: 92-106 (translated from Geokhimiya 1981, 2: 163-178)

Aubert G (1969) Les coupoles granitiques de Montebras et d'Échassières (Massif Central français) et la genèse de leurs minéralisations en étain, lithium, tungstène et béryllium. Mém BRGM 46: 1-346

Avila SW (1982) Casiterita esferulítica de Chacaltaya. CEDOMIN < La Paz> 5: 37-52

Ball TK, Basham IR (1984) Petrogenesis of the Bosworgey granitic cusp in the SW England tin province and its implications for ore mineral genesis. Mineral Deposita 19: 70-77

Banks NG, Cornwall HR, Silbermann ML, Creasey SC, Marvin RF (1972) Chronology of intrusion and ore deposition at Ray, Arizona: Part 1, K-Ar ages. Econ Geol 67: 864-878

Bard JP, Botello M, Martinez C, Subieta T (1974) Relations entre tectonique, métamorphisme et mise en place d'un granite éohercynien à deux micas dans la Cordillère Real de Bolivie (Massif de Zongo-Yani). Cah ORSTOM < Paris>, Sér Géol 6: 3-18

Barnes PBJr (1984) Redox reactions in hydrothermal fluids. In: Henley RW, Truesdell AH, Barton PBJr (eds) Fluid-mineral equilibria in hydrothermal systems. Rev Econ Geol 1: 99-113

Barosh PJ (1968) Relationship of explosion-produced fracture patterns to geologic structure in Yucca Flat, Nevada Test Site. Mem Geol Soc Am 110: 199-217

Barsukov VL (1956) New data on the geochemistry of tin. Abstr./Resum. Trabajos Presentados, 20th Int Geol Congr, Mexico-City, 355 pp

Barsukov VL (1957) The geochemistry of tin. Geochemistry 1: 41-52 (translated from Geokhimiya 1957: 36-45)

Baumann L (1965) Zur Erzführung und regionalen Verbreitung des "Felsithorizontes" von Halsbrücke. Freiberg Forschungsh C 186: 63-81

Baumann L, Tägl F (1963) Neue Erkundungsergebnisse zur Tektonik und Genesis der Zinnerzlagerstätte von Ehrenfriedersdorf. Freiberg Forschungsh C 167: 35-63

Baumann L, Weinhold G (1963) Zum Neuaufschluss des sogenannten "Felsithorizontes" von Halsbrücke. Z Angew Geol 9: 338-345

Baumann L, Tischendorf G (1976) Einführung in die Metallogenie/ Minerogenie. Dtsch Verlag Grundstoffind, Leipzig, 458 pp

Beaumont E de (1847) Note sur les émanations volcaniques et métallifères. Bull Soc Géol France, 2ième Sér, 4: 1249-1334

Beckinsale RD (1979) Granite magmatism in the tin belt of Southeast Asia. In: Atherton MP, Tarney J (eds) Origin of granite batholiths. Shiva, Orpington, pp 34-44

Beckinsale RD, Suensilpong S, Nakapadungrat S, Walsh JN (1979) Geochronology and geochemistry of granite magmatism in Thailand in relation to a plate tectonic model. J Geol Soc London 136: 529-540

Berg G (1922) Die Gesteine des Isergebirges. Jb Preuss Geol Landesanst 43: 125-168

Beyschlag F, Krusch P, Vogt JHL (1910) Die Lagerstätten der nutzbaren Mineralien und Gesteine nach Form, Inhalt und Entstehung. Enke, Stuttgart, Bd 1, 509 pp

Bignell JD, Snelling NJ (1977) Geochronology of Malayan granites. Overseas Geol Min Res, Inst Geol Sci London 47: 1-72

Biste M (1979) Die Anwendung geochemischer Indikatoren auf die Zinn-Höffigkeit herzynischer Granite in Süd-Sardinien. Berliner Geowiss Abh A 18: 1-107

Blatt H, Jones RL (1975) Proportions of exposed igneous, metamorphic, and sedimentary rocks. Bull Geol Soc Am 86: 1085-1088

Boissavy-Vinau M (1979) Processus géochimiques de concentrations liés à l'évolution de magmas granitiques. Application aux filons à Sn-W du Massif Central et du Nord Portugal. Thèse 3ième cycle, Univ Paris VI, Paris, 220 pp

Boissavy-Vinau M, Roger G (1980) The TiO_2/Ta ratio as an indicator of the degree of differentiation of tin granites. Mineral Deposita 15: 231-236

Bolduan H (1963) Geologie und Genese der Zinn-Wolframlagerstätte Geyer (Erzgebirge). Freiberg Forschungsh C 167: 7-34

Bourcier WL, Barnes HL (1987) Ore solution chemistry - VII. Stabilities of chloride and bisulfide complexes of zinc to 350 °C. Econ Geol 82: 1839-1863

Bowden P (1982) Magmatic evolution and mineralization in the Nigerian younger granite province. In: Evans AM (ed) Metallization associated with acid magmatism. Wiley, Chichester, pp 51-61

Bowden P, Kinnaird JA (1984) Geology and mineralization of the Nigerian anorogenic ring complexes. Geol Jb <Hannover> B 56: 3-65

Bräuer H (1970) Spurenelementgehalte in granitischen Gesteinen des Thüringer Waldes und des Erzgebirges. Freiberg Forschungsh C 259: 83-139

Breemen O van, Hutchinson J, Bowden P (1975) Age and origin of the Nigerian Mesozoic granites: a Rb-Sr isotopic study. Contrib Mineral Petrol 50: 157-172

Brewer MS, Lippolt HJ (1974) Petrogenesis of basement rocks of the upper Rhine region elucidated by rubidium-strontium systematics. Contrib Mineral Petrol 45: 123-141

Brimhall GH Jr (1987) Preliminary fractionation patterns of ore metals through Earth history. Chem Geol 64: 1-16

Brimhall GH Jr, Crerar DA (1987) Ore fluids: magmatic to supergene. Rev Mineral 17: 235-321

Brown GC, Thorpe RS, Webb PC (1984) The geochemical characteristics of granitoids in contrasting arcs and comments on magma sources. J Geol Soc London 141: 413-426

Brown GF, Buravas S, Charaljavanaphet J, Jalichandra N, Johnston WD Jr, Sresthaputra V, Taylor GC Jr (1951) Geologic reconnaissance of the mineral deposits of Thailand. US Geol Surv Bull 984: 1-183

Bryndzia LT, Wood BJ, Dick HJB (1989) The oxidation state of the Earth's sub-oceanic mantle from oxygen thermobarometry of abyssal spinel peridotites. Nature <London> 341: 526-527

Buch L von (1802) Geognostische Beobachtungen auf Reisen durch Deutschland und Italien. Erster Band. Haude-Spener, Berlin, 320 pp

Buchanan MS, MacLeod WN, Turner DC, Wright EP (1971) The geology of the Jos Plateau, Vol 2. The younger granite complexes. Nigeria Geol Surv Bull 32: 1-160

Budzinski H, Tischendorf G (1985) Modelling of fractional crystallization of granitic magmas: the Variscan postkinematic Older Granites of Western Erzgebirge-Vogtland, G.D.R. Gerlands Beitr Geophys 94: 370-380

Burnham CW (1967) Hydrothermal fluids at the magmatic stage. In: Barnes HL (ed) Geochemistry of hydrothermal ore deposits. Holt Rinehart Winstone, New York, pp 34-76

Burnham CW (1979a) Magmas and hydrothermal fluids. In: Barnes HL (ed) Geochemistry of hydrothermal ore deposits, 2nd edn. Wiley, New York, pp 71-136

Burnham CW (1979b) The importance of volatile constituents. In: Yoder HS (ed) The evolution of the igneous rocks. Fiftieth anniversary perspectives. Princeton Univ Press, Princeton, pp 439-482

Burnham CW (1985) Energy release in subvolcanic environments: implications for breccia formation. Econ Geol 80: 1515-1522

Burnham CW, Davis NF (1971) The role of H_2O in silicate melts. I. P-V-T relations in the system $NaAlSi_3O_8-H_2O$. Am J Sci 270: 54-79

Burnham CW, Ohmoto H (1980) Late-stage processes of felsic magmatism. Mining Geol, Spec Issue 8: 1-11

Burnol L (1974) Géochimie du béryllium et types de concentrations dans les leucogranites du Massif Central français. Relation entre les caractéristiques des granitoides et les gisements endogènes de type départ acide (Be, Sn, Li) ou de remaniement tardif (U, F, Pb, et Zn). Mém BRGM 85: 1-170

Burnol L (1978) Different types of leucogranites and classification of the types of mineralization associated with acid magmatism in the north-western part of the French Massif Central. In: Stemprok M, Burnol L, Tischendorf G (eds) Metallization associated with acid magmatism, Vol 3. Czech Geol Surv, Praha, pp 191-204

Burt DM, Sheridan MF, Bikun JV, Christiansen EH (1982) Topaz rhyolites - distribution, origin, and significance for exploration. Econ Geol 77: 1818-1836

Cabello J (1986) Precious metals and Cenozoic volcanism in the Chilean Andes. J Geochem Expl 25: 1-19

Cameron KL (1984) Bishop Tuff revisited: new rare-earth element data consistent with fractional crystallization. Science 224: 1338-1340

Candela PA, Holland HD (1984) The partitioning of copper and molybdenum between silicate melts and aqueous fluids. Geochim Cosmochim Acta 48: 373-380

Candela PA, Bouton SL (1990) The partitioning of tungsten and molybdenum between silicate melts and ilmenite: implications for fO_2 control of metal ratios in magmatic-hydrothermal W-Mo deposits. Econ Geol 85 (in press)

Cann JR (1982) Rayleigh fractionation with continuous removal of liquid. Earth Planet Sci Lett 60: 114-116

Carten RB, Walker BM, Geraghty EP, Gunow AJ (1988) Comparison of field-based studies of the Henderson porphyry molybdenum deposit, Colorado, with experimental and theoretical models of porphyry systems. In: Taylor RP, Strong DF (eds) Recent advances in the geology of granite-related mineral deposits. CIM (Canad Inst Min Metal), Spec Vol 39: 351-366

Carvalho P (1986) An introduction to the Neves Corvo copper mine, Portugal. In: Thadeu D (ed) Guidebook Iberian field conference, 11-22 April 1986, Portuguese section. Soc Geol Appl Mineral Deposits (SGA), Lisboa, pp 83-99

Cerny P (1982) Petrogenesis of granitic pegmatites. In: Cerny P (ed) Short course in granitic pegmatites in science and industry. Mineral Assoc Canada, Short Course Handbook 8: 405-461

Cerny P (1989) Characteristics of pegmatite deposits of tantalum. In: Möller P, Cerny P, Saupé F (eds) Lanthanides, tantalum and niobium. Mineralogy, geochemistry, characteristics of primary ore deposits, prospecting, processing and applications. Proceedings of a workshop in Berlin, November 1986. Springer, Berlin Heidelberg New York, pp 195-239

Cerny P, Meintzer RE, Anderson AJ (1985) Extreme fractionation in rare-element granitic pegmatites: selected examples of data and mechanisms. Can Mineral 23: 381-421

Chappell BW, White AJR (1974) Two contrasting granite types. Pacific Geol 8: 173-174

Charoy B (1986) The genesis of the Cornubian batholith (SW England): the example of the Carnmenellis pluton. J Petrol 27: 571-604

Charpentier JFW (1778) Mineralogische Geographie der Chursächsischen Lande. Crusius, Leipzig, 432 pp

Chen Guoda (1989) Tectonics of China. Pergamon, Oxford, 266 pp

Cheng Xianyao, Huang Youde, Yao Jingyan, We Mingchao, Peng Zhenan, (1984) The primitive concentration of tin elements and the model of double controlled tin deposits. Report Int Symp Geology of Tin Deposits, Nanning and Dachang, China, 27 Oct-8 Nov 1984. Chinese Acad Geol Sci/ESCAP Reg Miner Res Develop Cent, Bandung, pp 33-34

Christiansen EH, Burt DM, Sheridan MF, Wilson RT (1983) The petrogenesis of topaz-rhyolites from the western United States. Contrib Mineral Petrol 83: 16-30

Clark AH, Farrar E, Caelles JC, Haynes SJ, Lortie RB, McBride SL, Quirt GS, Robertson RCR, Zentilli M (1976) Longitudinal variations in the metallogenetic evolution of the Central Andes: a progress report. Geol Assoc Canada, Spec Pap 14: 23-58

Clark AH, Robertson RCR (1978) The evolution and origin of the northern plutonic subprovinces of the Bolivian tin belt. Abstr, Int Symp Geology of Tin Deposits, March 1978, Kuala Lumpur, Annex to Warta Geol 4: 42-43

Cobbing EJ, Pitfield PEJ, Beckinsale RD (1988) Report on the geology and geochemistry of a selection of granites from Burma. NERC, British Geol Surv, Report WC/88/30/R, pp 1-45

Cobbing EJ (1989) Summary of petrographic notes and geochemical data on granite samples from Malaysia. British Geol Surv, Intern Report, 32 p

Cobbing EJ, Mallick DIJ, Pitfield PEJ, Teoh LH (1986) The granites of the Southeast Asian tin belt. J Geol Soc London 143: 537-550

Collins WJ, Beams SD, White AJR, Chappell BW (1982) Nature and origin of A-type granites with particular reference to southeastern Australia. Contrib Mineral Petrol 80: 189-200

Compston W, Chappell BW (1979) Sr-isotope evolution of granitoid source rocks. In: McElhinny MW (ed) The Earth: its origin, structure and evolution. Academic Press, London, pp 377-426

Cotta B von (1859) Die Lehre von den Erzlagerstätten. Erster Theil, 2. Aufl. Engelhardt, Freiberg, 252 pp

Crerar DA, Barnes HL (1976) Ore solution chemistry - V. Solubilities of chalcopyrite and chalcocite assemblages in hydrothermal solution at 200 to 350 °C. Econ Geol 71: 772-794

Criss RE, Taylor HP Jr (1986) Meteoric-hydrothermal systems. Rev Mineral 16: 373-424

Crowson P (1984) Minerals handbook 1984-85. Stockton, New York, 294 pp

Crowson P (1986) Minerals handbook 1986-87. Stockton, New York, 331 pp

Dadze TP, Sorokhin VI, Nekrasov IY (1981) Solubility of SnO_2 in water and in aqueous solutions of HCl, HCl+KCl, and HNO_3 at 200-400 °C and 101.3 MPa. Geochem Intern 18, 5: 142-152 (translated from Geokhimiya 1981, 10: 1482-1492)

Dahm KP, Gerstenberger H, Geissler M (1985) Zum Problem der Granitgenese im Erzgebirge, DDR. Z Geol Wiss 13: 545-557

Darbyshire DPF (1988a) Geochronology of Malaysian granites. NERC Isotope Geol Cent Report 88/3: 1-59

Darbyshire DPF (1988b) Geochronology of Tin Islands granites, Indonesia. NERC Isotope Geol Cent Report 88/4: 1-30

Darbyshire DPF (1988c) Geochronology of Thai granites. NERC Isotope Geol Cent Report 88/5: 1-44

Darbyshire DPF, Shepherd TJ (1985) Chronology of granite magmatism and associated mineralization, SW England. J Geol Soc London 142: 1159-1177

Darbyshire DPF, Shepherd TJ (1988) A Rb-Sr and Sm-Nd study of granites and related Sn-W mineralization in SW England. Abstr, 5th Int Symp Tin/Tungsten Granites in SE Asia and the western Pacific, Oct 17-19, 1988. Shimane Univ, Matsue, pp 30-31

Darbyshire DPF, Swainbank IG (1988) Geochronology of a selection of granites from Burma. NERC Isotope Geol Cent Report 88/6: 1-38

Darnley AG (1986) High heat production (HHP) granites, hydrothermal circulation and ore genesis. Trans Instn Min Metall 95: B46-B50

Daubrée A (1841) Mémoire sur le gisement, la constitution et l'origine des amas de minerai d'étain. Ann Mines, 3ième Sér 20: 65-112

Daubrée A (1849) Recherches sur la production artificielle de quelques espèces minérales cristallines, particulièrement de l'oxyde d'étain, de l'oxyde de titane et du quartz: observations sur l'origine des filons titanifères des Alpes. Comptes Rendus Hebd Séances Acad Sci <Paris> 29: 227-232

DePaolo DJ (1981) Trace element and isotopic effects of combined wallrock assimilation and fractional crystallization. Earth Planet Sci Lett 53: 189-202

DePaolo DJ (1988) Neodymium isotope geochemistry. An introduction. Springer, Berlin Heidelberg New York Tokyo, 187 pp

Dillet B, Czamanske GK (1987) Aspects of the petrology, mineralogy, and geochemistry of the granitic rocks associated with Questa Caldera, northern New Mexico. US Geol Surv, Open-File Rep 87-258: 1-238

Dingwell DB (1988) The structures and properties of fluorine-rich magmas: a review of experimental studies. In: Taylor RP, Strong DF (eds), Recent advances in the geology of granite-related mineral deposits. Canad Inst Min Metall, Spec Vol 39: 1-12

Dingwell DB, Harris DM, Scarfe CM (1984) The solubility of H_2O in melts in the system SiO_2-Al_2O_3-Na_2O-K_2O at 1 to 2 kbars. J Geol 92: 387-395

Dmitriev L, Barsukov V, Udintsev G (1971) Rift-zones of the ocean and the problem of ore-formation. Min Geol <Japan>, Spec Issue 3 (Proc IMA-IAGOD Meetings 1970), pp 65-69

Doerner HA, Hoskins WM (1925) Co-precipitation of radium and Ba sulfates. Am Chem Soc 47: 662-675

Drach V von, Lippolt HJ, Brewer M (1974) Rb-Sr-Altersbestimmungen an Graniten des Nordschwarzwaldes. N Jb Miner Abh 123: 38-62

Durasova NA, Barsukov VL, Ryabchikov IO, Khramov DA, Kravtsova RP (1984) The valency states of tin in basalts at various oxygen fugacities. Geochem Intern 21, 4: 7-8 (translated from Geokhimiya 1984, 3: 435-437)

Durisova J, Charoy B, Weisbrod A (1979) Fluid inclusion studies in minerals from tin and tungsten deposits in the Krusné Hory Mountains (Czechoslovakia). Bull Minéral 102: 665-675

Duthou JL (1978) Les granitoides du Haut Limousin (Massif central français); chronologie Rb-Sr de leur mise en place; le thermo-métamorphisme carbonifère. Bull Soc Géol France 20: 229-235

Eadington PJ (1988) The solubility of cassiterite in hydrothermal solutions in relation to some lithological and mineral associations of tin ores. In: Taylor RP, Strong DF (eds) Recent advances in the geology of granite-related mineral deposits. CIM (Can Inst Min Metal), Spec Vol 39: 25-32

Eastoe CJ (1982) Physics and chemistry of the hydrothermal system at the Panguna porphyry copper deposit, Bougainville, Papua New Guinea. Econ Geol 77: 127-153

El Bouseily AM, El Sokkary AA (1975) The relation between Rb, Ba and Sr in granitic rocks. Chem Geol 16: 207-219

Emmermann R (1977) A petrogenetic model for the origin and evolution of the Hercynian granite series of the Schwarzwald. N Jb Miner Abh 128: 219-253

Eugster HP (1985) Granites and hydrothermal ore deposits: a geochemical framework. Mineral Mag 49: 7-23

Eugster HP (1986) Minerals in hot water. Am Mineral 71: 655-673

Eugster HP, Wilson GA (1985) Transport and deposition of ore-forming elements in hydrothermal systems associated with granites. In: Halls C (ed) High heat production granites, hydrothermal circulation and ore genesis. Inst Min Metall, London, pp 87-98

Eugster HP, Wones DR (1962) Stability relations of the ferrugineous biotite, annite. J Petrol 3: 82-125

Ewart A (1981) The mineralogy and chemistry of the anorogenic Tertiary silicic volcanics of SE Queensland and NE New South Wales, Australia. J Geophys Res 86: 10242-10256

Exley CS, Stone M (1982) Petrogenesis. In: Sutherland DS (ed) Igneous rocks of the British Isles. Wiley, Chichester, pp 311-320

Farmer GL, DePaolo DJ (1983) Origin of Mesozoic and Tertiary granite in the western United States and implications for pre-Mesozoic crustal structure. 1. Nd and Sr isotopic studies in the geocline of the northern Great Basin. J Geophys Res 88: 3379-3401

Farmer GL, DePaolo DJ (1984) Origin of Mesozoic and Terrtiary granite in the western US and implications for pre-Mesozoic crustal structure. 2. Nd and Sr isotopic studies of unmineralized and Cu- and Mo-mineralized granite in the Precambrian craton. J Geophys Res 89: 10141-10160

Fehn U (1985) Post-magmatic convection related to high heat production in granites of southwest England: a theoretical study. In: Halls C (ed) High heat production (HHP) granites, hydrothermal circulation and ore genesis. Instn Min Metall, London, pp. 99-112

Fehn U, Cathles LM, Holland HD (1978) Hydrothermal convection and uranium deposits in abnormally radioactive plutons. Econ Geol 73: 1556-1566

Ferguson HG, Bateman AM (1912) Geologic features of tin deposits. Econ Geol 7: 209-262

Fernandez A, Hörmann PK, Kussmaul S, Meave J, Pichler H, Subieta T (1973) First petrologic data on young volcanic rocks of SW-Bolivia. Tschermaks Min Petr Mitt 19: 149-172

Feyerabend P (1979) Contre la méthode. Esquisse d'une théorie anarchiste de la connaissance. Seuil, Paris, 350 pp

Fleischer R, Routhier P (1970) Quelques grands thèmes de la géologie du Brésil. Miscellanées géologiques et métallogéniques sur le Planalto. Sci Terres 15: 45-102

Fletcher WK, Dousset PE, Yusoff I (1984) Bahaviour of tin and associated elements in a mountain stream, Bujang Melaka, Perak, Malaysia. SEATRAD Centre <Ipoh, Malaysia>, Report Invest 24: 1-71

Floyd PA, Exley CS, Stone M (1983) Variscan magmatism in southwest England - discussion and synthesis. In: Hancock PL (ed) The Variscan fold belt in the British Isles. Hilger, Bristol, pp 178-185

Förster HJ, Tischendorf G (1989) Reconstruction of the volatile characteristics of granitoidic magmas and hydrothermal solutions on the basis of dark micas: The Hercynian postkinematic granites and associated high-temperature mineralizations of the Erzgebirge (G.D.R.). I. Communication: Calculation procedure and results. Chem Erde 49: 7-20

Förster HJ, Tischendorf G (1990) Reconstruction of the volatile characteristics of granitoidic magmas and hydrothermal solutions on the basis of dark micas: The Hercynian postkinematic granites and associated high-temperature mineralizations of the Erzgebirge (G.D.R.). II. Communication: Implications to metallogenesis. Chem Erde, in press.

Fox W (1969) Tin mining in Spain and Portugal: a paper of information. In: Fox W (ed) A second technical conference on tin. Int Tin Council, Bangkok London, pp 223-265

French BM, Eugster HP (1965) Experimental control of oxygen fugacities by graphite-gas equilibrium. J Geophys Res 70: 1529-1539

Gardeweg M, Ishihara S, Matsuhisa Y, Shibata K, Terashima S (1984) Geochemical studies of Upper Cenozoic igneous rocks from the Altiplano of Anto-fagasta, Chile. Bull Geol Surv Japan 35: 547-563

Gaupp R, Möller P, Morteani G (1984) Tantal-Pegmatite. Geologische, petrologische und geochemische Untersuchungen. Monogr Ser Mineral Deposits 23: 1-124

Gerstenberger H (1989) Autometasomatic Rb enrichments in highly evolved granites causing lowered Rb-Sr isochron intercepts. Earth Planet Sci Lett 93: 65-75

Gerstenberger H, Haase G, Tischendorf G, Wetzel K (1984) Zur Genese der variszisch-postkinematischen Granite des Erzgebirges. Chem Erde 43: 263-277

Giggenbach WF (1980) Geothermal gas equilibria. Geochim Cosmochim Acta 44: 2021-2032

Gill JB (1978) Role of trace element partition coefficients in models of andesite genesis. Geochim Cosmochim Acta 42: 709-724

Goethe JW (1785,1963) Zur Theorie der Gesteinslagerung. In: Johann Wolfgang Goethe - dtv-Gesamtausgabe, Bd 38. Dtsch Taschenbuch Verlag, pp 15-16

Goldschmidt VM (1937) The principles of distribution of chemical elements in minerals and rocks. J Chem Soc <London> 1937: 655-672

Grant JN, Halls C, Avila W, Avila G (1977) Igneous geology and the evolution of hydrothermal systems in some sub-volcanic tin deposits of Bolivia. Geol Soc London, Spec Publ 7: 117-126

Grant JN, Halls C, Avila W, Snelling NJ (1979) K-Ar ages of igneous rocks and mineralization in part of the Bolivian tin belt. Econ Geol 74: 838-851

Grant JN, Halls C, Sheppard SMF, Avila W (1980) Evolution of the porphyry tin deposits of Bolivia. Min Geol, Spec Issue 8: 151-173

Grohmann H (1965) Beitrag zur Geochemie österreichischer Granitoide. Tschermaks Min Petr Mitt 10: 436-474

Groves DI (1972) The geochemical evolution of tin-bearing granites in the Blue Tier batholith, Tasmania. Econ Geol 67: 445-457

Groves DI, McCarthy TS (1978) Fractional crystallization and the origin of tin deposits in granitoids. Mineral Deposita 13: 11-26

Guan Xunfan, Zhou Yongqin, Xiao Jinghua, Liang Shuzhao, Li Jinmao (1984) A new type of tin deposits in China - the Yinyan porphyry tin deposit. Report on the international symposium on the geology of tin deposits (Nanning and Dachang, 27 Oct-8 Nov, 1984). Chinese Acad Geol Sci/ESCAP Regional Min Res Develop Progr, Bandung, pp 40

Gustafson LB, Hunt JP (1975) The porphyry copper deposit at El Salvador, Chile. Econ Geol 70: 857-912

Haffty J, Noble DC (1972) Release and migration of molybdenum during the primary crystallization of peralkaline silicic volcanic rocks. Econ Geol 67: 768-775

Hall A (1971) Greisenisation in the granite of Cligga Head, Cornwall. Proc Geol Assoc 82: 209-230

Hall A (1974) Granite porphyries in Cornwall. Proc Ussher Soc 3: 145-149

Hall A (1990) Geochemistry of the Cornubian tin province. Mineral Deposita 25: 1-6

Halliday AN (1980) The timing of early and main stage ore mineralization in southwest Cornwall. Econ Geol 75: 752-759

Hamaguchi H, Kuroda R, Onuma N, Kawabuchi K, Mitsubayashi T, Hosohara K (1964) The geochemistry of tin. Geochim Cosmochim Acta 28: 1039-1053

Hamaguchi H, Kuroda R (1969) Tin. In: Wedepohl KH (ed) Handbook of geochemistry, Vol. II/4. Springer, Berlin Heidelberg, 50-B-1 to 50-M-5

Hanus D (1982) The Colquiri tin deposit: a contribution to its genesis. In: Amstutz GC, El Goresy A, Frenzel G, Kluth C, Moh G, Wauschkuhn A, Zimmermann RA (eds) Ore genesis: the state of the art. Springer, Berlin Heidelberg New York Tokyo, pp 308-316

Hards N (1978) Distribution of elements between the fluid phase and silicate melt phase of granites and nepheline syenites. NERC Prog Exp Petr 5 (4): 88-90

Haselton HT Jr, D'Angelo WM (1986) Tin and tungsten solubilities (500-700 C, 1 kbar) in the presence of a synthetic quartz monzonite. Eos 67: 388

Hawkesworth CJ, Hammill M, Gledhill AR, Calsteren P van, Rogers G (1982) Isotope and trace element evidence for late-stage intra-crustal melting in the High Andes. Earth Planet Sci Lett 58: 240-254

Haynes FM, Titley SR (1980) The evolution of fracture-related permeability within the Ruby Star Granodiorite, Sierrita porphyry copper deposit, Pima County, Arizona. Econ Geol 75: 673-683

Hegel GWF (1807) System der Wissenschaft. Erster Theil, die Phänomenologie des Geistes. Goebhardt, Bamberg, 767 pp

Heidrick TL, Titley SR (1982) Fracture and dike patterns in Laramide plutons and their structural and tectonic implications. In: Titley SR (ed) Advances in geology of the porphyry copper deposits. Univ. Arizona Press, Tucson, pp 73-91

Heinrich CA (1990) The chemistry of hydrothermal tin-tungsten ore deposition. Mscr (submitted to Econ Geol), 55 pp

Heinrich CA, Eadington PJ (1986) Thermodynamic predictions of the hydrothermal chemistry of arsenic, and their significance for the paragenetic sequence of some cassiterite-arsenopyrite-base metal sulfide deposits. Econ Geol 81: 511-529

Heinrich CA, Jaireth S (1989) Fluid buffering and wall rock interaction in the chemical evolution of granite-related tin-tungsten veins. 28th Int Geol Congr Washington, Abstr 2: 47-48

Helgeson HC, Delaney JM, Nesbitt HW, Bird DK (1978) Summary and critique of the thermodynamic properties of rock-forming minerals. Am J Sci 278-A: 1-229

Helmcke D (1985) The Permo-Triassic "Paleotethys" in mainland Southeast-Asia and adjacent parts of China. Geol Rdsch 74: 215-228

Hemley JJ, Jones WR (1964) Chemical aspects of hydrothermal alteration with emphasis on hydrogen metasomatism. Econ Geol 59: 538-569

Herzenberg R (1936) Colloidal tin ore deposits. Econ Geol 31: 761-766

Hewitt DA (1978) A redetermination of the fayalite-magnetite-quartz equilibrium between 650 and 850 °C. Am J Sci 278: 715-724

Higgins NC, Sun SS (1988) Radiogenic tracers of hydro-thermal fluids in tin-tungsten ore systems. Extended Abstr, 5th Int Symp Tin/Tungsten Granites in Southeast Asia and the Western Pacific, Oct 17-19, 1988. Shimane Univ, Matsue, pp 43-47

Higgins NC, Solomon M, Varne R (1985) The genesis of the Blue Tier batholith, northeastern Tasmania, Australia. Lithos 18: 129-149

Higgins NC, Forsythe DL, Sun SS, Andrew AS (1987) Fluid and metal sources in the Mt Carbine tungsten deposit, North Queensland, Australia. Proc Pacific Rim Congress, 26-29 Aug 1987. Australasian Inst Metall, Gold Coast, pp 173-177

Hildreth W (1979) The Bishop Tuff: evidence for the origin of compositional zonation in silicic magma chambers. Geol Soc Am Spec Pap 180: 43-75

Hildreth W (1981) Gradients in silicic magma chambers: implications for lithospheric magmatism. J Geophys Res 86: 10153-10192

Hildreth W, Moorbath S (1988) Crustal contributions to arc magmatism in the Andes of Central Chile. Contrib Mineral Petrol 98: 455-489

Hine R, Williams IS, Chappell BW, White AJR (1978) Contrasts between I- and S-type granitoids of the Kosciusko batholith. J Geol Soc Australia 25: 219-234

Hirschwald W, Knacke O, Reinitzer P (1957) Thermodynamische Daten und Gleichgewichtsdiagramme metallurgischer Systeme. Erzmetall 10: 123-127

Holden P, Halliday AN, Stephens WE (1987) Neodymium and strontium isotope content of microdiorite enclaves points to mantle input to granitoid production. Nature <London> 330: 53-56

Holder MT, Leveridge BE (1986) A model for the tectonic evolution of south Cornwall. J Geol Soc London 143: 125-134

Holland HD (1972) Granites, solutions, and base metal deposits. Econ Geol 67: 281-301

Holloway JR (1976) Fluids in the evolution of granitic magmas - consequences of finite CO_2 solubility. Geol Soc Am Bull 87: 1513-1518

Hudson T, Arth JG (1983) Tin granites of Seward Peninsula, Alaska. Geol Soc Am Bull 94: 768-790

Hudson T, Smith JG, Elliott RL (1979) Petrology, composition, and age of intrusive rocks associated with the Quartz Hill molybdenite deposit, southeastern Alaska. Can J Earth Sci 16: 1805-1822

Hudson T, Arth JG, Muth KG (1981) Geochemistry of intrusive rocks associated with molybdenite deposits, Ketchikan Quadrangle, southeastern Alaska. Econ Geol 76: 1225-1232

Huebner JS, Sato M (1970) The oxygen fugacity-temperature relationships of manganese oxide and nickel oxide buffers. Am Mineral 55: 934-952

Humboldt A von (1823a) Essai géognostique sur le gisement des roches dans les deux hémisphères. Levrault, Strasbourg, 364 pp

Humboldt A von (1823b) Geognostischer Versuch über die Lagerung der Gebirgsarten in beiden Erdhälften. Deutsch bearbeitet von Karl Cäsar Ritter von Leonhard. Levrault, Straßburg, 383 pp

Hunter DR (1973) The localization of tin mineralization with reference to southern Africa. Minerals Sci Engng 5: 53-77

Hurley PM, Rand JR (1969) Pre-drift continental nuclei. Science 164: 1229-1242

Huspeni JR, Kesler SE, Ruiz J, Tuta Z, Sutter JF, Jones LM (1984) Petrology and geochemistry of rhyolites associated with tin mineralization in northern Mexico. Econ Geol 79: 87-105

Hutchison CS (1983) Multiple Mesozoic Sn-W-Sb granitoids of southeast Asia. Geol Soc Amer Mem 159: 35-60

Imeokparia EG (1980) Ore-bearing potential of granitic rocks from the Jos-Bukuru Complex, northern Nigeria. Chem Geol 28: 69-77

Imeokparia EG (1984) Geochemistry of the granitic rocks from the Kwandonkaya Complex, northern Nigeria. Lithos 17: 103-115

Imeokparia EG (1986a) Geochemical evolution of the metaluminous and peraluminous granites of Ganawuri Younger Granite Complex, northern Nigeria. J Afr Earth Sci 5: 193-200

Imeokparia EG (1986b) The geochemistry and petrogenesis of rocks of the Buji Younger Granite Complex, northern Nigeria. Chem Erde 45: 301-320

Ingham FT, Bradford EF (1960) The geology and mineral resources of the Kinta Valley, Perak. Geol Surv Malaysia, District Mem 9: 1-347

Isaacson PE (1975) Evidence for a western extracontinental land source during the Devonian period in the Central Andes. Geol Soc Am Bull 86: 39-46

Ishihara S (1967) Molybdenum mineralization at Questa Mine, New Mexico, USA. Geol Surv Japan Report 218: 1-64

Ishihara S (1977) The magnetite-series and ilmenite-series granitic rocks. Min Geol 27: 293-305

Ishihara S (1981) The granitoid series and mineralization. Econ Geol, 75th Anniv Vol, pp 458-484

Ishihara S, Sawata H, Arpornsuwan S, Busaracome P, Bungbrakearti N (1979) The magnetite-series and ilmenite-series granitoids and their bearing on tin mineralization, particularly of the Malay Peninsula region. Geol Soc Malaysia Bull 11: 103-110

Ishihara S, Ulriksen CE, Sato K, Terashima S, Sato T, Endo Y (1984) Plutonic rocks of north-central Chile. Geol Surv Japan Bull 35: 503-536

ITC (1967) Statistical yearbook 1966. Int Tin Counc, The Hague, 281 pp

ITRDC (1938) Statistical yearbook 1938. Int Tin Res Develop Counc, The Hague, 206 pp

Jackson KJ, Helgeson HC (1985a) Chemical and thermodynamic constraints on the hydrothermal transport and deposition of tin: I. Calculation of the solubility of cassiterite at high pressures and temperatures. Geochim Cosmochim Acta 49: 1-22

Jackson KJ, Helgeson HC (1985b) Chemical and thermodynamic constraints on the hydrothermal transport and deposition of tin: II. Interpretation of phase relations in the Southeast Asian tin belt. Econ Geol 80: 1365-1378

Jackson NJ, Halliday AN, Sheppard SMF, Mitchell JG (1982) Hydrothermal activity in the St Just mining district, Cornwall, England. In: Evans AM (ed) Mineralization associated with acid magmatism. Wiley, New York, pp 137-179

Jacobsen SB (1988) Isotopic and chemical constraints on mantle-crust evolution. Geochim Cosmochim Acta 52: 1341-1350

Jacobson HS, Pierson CT, Danusawad T, Japakasetr T, Ithuputi B, Sirirat-anamaongkol C, Prapassornkul S, Pholphan N (1969) Mineral investigations in northeastern Thailand. US Geol Surv Prof Pap 618: 1-96

Jahns RH, Burnham CW (1969) Experimental studies of pegmatite genesis: I. A model for the derivation and crystallization of granitic pegmatites. Econ Geol 64: 843-864

James RS, Hamilton DL (1969) Phase relations in the system $NaAlSi_3O_8$-$KAlSi_3O_8$-$CaAl_2Si_2O_8$-SiO_2 at 1 kb water vapour pressure. Contrib Miner Petr 21: 111-141

Jankovic S (1972) The origin of base-metal mineralization on the mid-Atlantic ridge (based upon the pattern of Iceland). Abstr, 24th Int Geol Congr Montreal, Sect 4: 326-334

Jaskolski S (1960) Beitrag zur Kenntnis über die Herkunft der Zinnlagerstätten von Gierczyn (Giehren) im Iser-Gebirge, Niederschlesien. N Jb Miner Abh 94: 181-190

Jaskolski S (1962) Erwägungen über die Genese zinnführender Schiefer im Isergebirge (Niederschlesien). Polska Akad Nauk, Prace Geol 12: 33-53

Johnson CM, Lipman PW (1988) Origin of metaluminous and alkaline volcanic rocks of the Latir volcanic field, northern Rio Grande rift, New Mexico. Contrib Mineral Petrol 100: 107-128

Johnston WD (1965) Oxidation-reduction equilibria in molten $Na_2O \cdot 2SiO_2$ glass. J Amer Ceramic Soc 48: 184-190

Jones MT, Reed BL, Doe BR, Lanphere MA (1977) Age of tin mineralization and plumbotectonics, Belitung, Indonesia. Econ Geol 72: 745-752

Jones WR, Hernon RM, Moore SL (1967) General geology of Santa Rita Quadrangle, Grant County, New Mexico. US Geol Surv Prof Pap 555: 1-144

Just G, Schilka W, Seltmann R (1987) INAA investigations in tin-bearing granites of the Altenberg and Sadisdorf ore deposits. 4th Meeting on Nuclear Analytical Methods, Dresden May 4-8, 1987, Proc 1: 242-251

Karsten DLG (1806) Über das Alter der Metalle. Eine Vorlesung gehalten in der öffentlichen Sitzung der Philomatischen Gesellschaft zu Berlin am 3ten April 1806. Gilberts Annal Physik 1806 (5): 1-21

Keith JD, Shanks III WC (1988) Chemical evolution and volatile fugacities of the Pine Grove porphyry molybdenum and ash-flow tuff system, southwestern Utah. In: Taylor RP, Strong DF (eds) Recent advances in the geology of granite-related mineral deposits. Can Inst Min Metal, Spec Vol 39: 402-423

Keith SB (1984) Magma series and mineral deposits. MagmaChem, Phoenix, 111 pp

Kelly WC, Rye RO (1979) Geologic, fluid inclusion, and stable isotope studies of the tin-tungsten deposits of Panasqueira, Portugal. Econ Geol 74: 1721-1822

Kelly WC, Turneaure FS (1970) Mineralogy, paragenesis and geothermometry of the tin and tungsten deposits of the Eastern Andes, Bolivia. Econ Geol 65: 609-680

Kinnaird JA (1985) Hydrothermal alteration and mineralization of the alkaline anorogenic ring complexes of Nigeria. J Afr Earth Sci 3: 229-251

Kilinc IA, Burnham CW (1972) Partitioning of chloride between a silicate melt and coexisting aqueous phase from 2 to 8 kilobars. Econ Geol 67: 231-235

Kilinc A, Carmichael ISE, Rivers ML, Sack RO (1983) The ferric-ferrous ratio of natural silicate liquids equilibrated in air. Contrib Mineral Petrol 83: 136-140

Klerkx J, Deutsch S, Pichler H, Zeil W (1977) Strontium isotopic composition and trace element data bearing on the origin of Cenozoic volcanic rocks of the Central and Southern Andes. J Volcanol Geotherm Res 2: 49-71

Klintsova AP, Barsukov VL (1973) Solubility of cassiterite in water and in aqueous NaOH solutions at elevated temperatures. Geochem Intern 1973: 540-547 (translated from Geokhimiya 1973, 5: 701-709)

Kolbe P (1966) Geochemical investigation of the Cape Granite, south-western Cape Province, South Africa. Transact Geol Soc S Afr 69: 161-199

Kolbe P, Taylor SR (1966a) Geochemical investigation of the granitic rocks of the Snowy Mountains area, New South Wales. J Geol Soc Australia 13: 1-25

Kolbe P, Taylor SR (1966b) Major and trace element relationships in granodiorites and granites from Australia and South Africa. Contrib Mineral Petrol 12: 202-222

Kontak DJ, Clark AH, Farrar E, Strong DF (1984) The rift-associated Permo-Triassic magmatism of the Eastern Cordillera: a precursor to the Andean orogeny. In: Pitcher WS, Atherton MP, Cobbing EJ, Beckinsale RD (eds) Magmatism at a plate edge: the Peruvian Andes. Blackie, London, pp 36-44

Kormilicyn VS (1987) Über die Genese und die metallo-genetische Bedeutung der Sulfid-Kassiterit-Vererzung des "Felsithorizonts" im Erzgebirge (DDR). Z Geol Wiss 15: 599-618

Kovalenko VI, Ryabchikov ID, Bogatikov OA (1984) Problems of ore-mineralization with acidic magmatism. In: Proceedings of the Sixth Quadrennial IAGOD Symposium. Schweizerbart, Stuttgart, pp 369-374

Kovalenko VI, Ryzhenko BN, Barsukov VL, Klintsova AP, Velyukhanova TK, Volynets MP, Kitayeva LP (1986) The solubility of cassiterite in HCl and HCl + NaCl (KCl) solutions at 500 °C and 1000 atm under fixed redox conditions. Geochem Intern 23, 7: 1-16 (translated from Geokhimiya 1986, 2: 190-205)

Kovalenko VI, Ryabchikov ID, Antipin VS (1988) Temperature dependence of the distribution coefficients for Sn, W, Pb, and Zn in magmatic systems. Geochem Intern 25, 1: 1-10 (translated from Geokhimiya 1987, 6: 755-764)

Kozlowski A (1978) Pneumatolytic and hydrothermal activity in the Karkonosze-Izera block. Acta Geol Polon 28: 171-222

Kozlowski A, Karwowski L (1975) Genetyczne wskazniki mineralizacji W-Sn-Mo na obszarze karkonosko-izerskim (with English summary). Kwart Geol 19(1): 67-73

Kozlowski A, Karwowski L, Olszynski W (1975) Tungsten-tin-molybdenum mineralization in the Karkonosze massif. Acta Geol Polon 25: 415-430

Kress VC, Carmichael ISE (1988) Stoichiometry of the iron oxidation reaction in silicate melts. Am Mineral 73: 1267-1274

Kudrin AV (1989) Behavior of Mo in aqueous NaCl and KCl solutions at 300-450 C. Geochem Intern 26, 8: 87-99 (translated from Geokhimiya 1989, 1: 99-112)

Kussmaul S, Hörmann PK, Ploskonka E, Subieta T (1977) Volcanism and structure of southwestern Bolivia. J Volcanol Geotherm Res 2: 73-111

Kwak TAP (1987) W-Sn skarn deposits and related metamorphic skarns and granitoids. Elsevier, Amsterdam, 451 pp

Laffitte P, Permingeat F, Routhier P (1965) Cartographie métallogénique, métallotecte, géochimie régionale. Bull Soc Franç Minér Crist 88: 3-6

Lange H, Tischendorf G, Pälchen W, Klemm I, Ossenkopf W (1972) Fortschritte der Metallogenie im Erzgebirge. B. Zur Petrographie und Geochemie der Granite des Erzgebirges. Geologie 21: 457-489

Langmuir CH (1989) Geochemical consequences of in situ crystallization. Nature <London> 340: 199-205

Lanier G, John EC, Swensen AJ, Reid J, Bard CE, Caddey SW, Wilson JC (1978) General geology of the Bingham Mine, Bingham Canyon, Utah. Econ Geol 77: 50-59

Launay L de (1913) Traité de métallogénie. Gîtes minéraux et métallifères. Tome 2. Béranger, Paris, 801 pp

Lebedev LM (1967) Metacolloids in endogenic deposits. Plenum, New York, 298 pp

Lehmann B (1979) Schichtgebundene Sn-Lagerstätten in der Cordillera Real/Bolivien. Berliner Geowiss Abh A 14: 1-135

Lehmann B (1982) Metallogeny of tin: magmatic differentiation versus geochemical heritage. Econ Geol 77: 50-59

Lehmann B (1985) Formation of the strata-bound Kellhuani tin deposits, Bolivia. Mineral Deposita 20: 169-176

Lehmann B (1987) Tin granites, geochemical heritage, magmatic differentiation. Geol Rdsch 76: 177-185

Lehmann B (1988a) Tin-bearing and tin-barren granites in Central Thailand. Report on the project "Tin-bearing and tin-barren granites, primary tin mineralization in Thailand", Vol 1 (Subproject Central Thailand). Hannover, Bundesanst Geowiss Rohstoffe, 220 pp

Lehmann B (1988b) Tanjungpandan Pluton, Belitung Island (Indonesia). Report on the project "Tin-bearing and tin-barren granites, primary tin mineralization in Indonesia, Vol. 4 (Subproject Tanjungpandan, Belitung). Hannover, Bundesanst Geowiss Rohstoffe, 155 pp

Lehmann B, Harmanto (1990) Large-scale tin depletion in the Tanjungpandan tin granite, Belitung Island, Indonesia. Econ Geol 85: 99-111

Lehmann B, Lavreau J (1987) Tin granites of the northern Kibaran belt, Central Africa (Kivu/Zaire, Rwanda, Burundi). In: Matheis G, Schandelmeier

H (eds) Current research in African earth sciences. Balkema, Rotterdam, pp 33-36

Lehmann B, Lavreau J (1988) Geochemistry of tin granites from Kivu (Zaire), Rwanda and Burundi. IGCP 255 Newslett 1: 43-46

Lehmann B, Mahawat C (1989) Metallogeny of tin in central Thailand: a genetic concept. Geology 17: 426-429

Lehmann B, Pichler H (1980) Tin distribution in mid-Andean volcanic rocks. Mineral Deposita 15: 35-39

Lehmann B, Schneider HJ (1981) Strata-bound tin deposits. In: Wolf KH (ed) Handbook of strata-bound and stratiform ore deposits, Vol 9. Elsevier, Amsterdam, pp 743-771

Lehmann B, Petersen U, Santivañez R, Winkelmann L (1988) Distribución geoquímica de estaño y boro en la secuencia paleozoica inferior de la Cordillera Real de Bolivia. Bol Soc Geol Perú 77: 19-27

Lehmann JG (1751) Kurtze Einleitung in einige Theile der Bergwercks-Wissenschaft. Anfängern zum Besten. Nicolai, Berlin, 192 pp

Lehmann JG (1753) Abhandlung von den Metall-Müttern und der Erzeugung der Metalle aus der Naturlehre und Bergwerckswissenschaft hergeleitet und mit chymischen Versuchen erwiesen. Nicolai, Berlin, 268 pp

Le Maitre RW (1976) The chemical variability of some common igneous rocks. J Petrol 17: 589-637

Lenthall DH, Hunter DR (1977) The geochemistry of the Bushveld granites in the Potgietersrus tin-field. Precambrian Res 5: 359-400

Leva M (1959) Fluidization. McGraw-Hill, New York, 327 pp

Liew TC (1983) Petrogenesis of the peninsular Malaysian granitoid batholiths. Unpubl Ph.D. thesis, Austr Nat Univ, Canberra, 291 pp

Liew TC, McCulloch MT (1985) Genesis of granitoid batholiths of Peninsular Malaysia and implications for models of crustal evolution: evidence from a Nd-Sr isotopic and U-Pb zircon study. Geochim Cosmochim Acta 49: 587-600

Liew TC, Page RW (1985) U-Pb zircon dating of granitoid plutons from the West Coast province of peninsular Malaysia. J Geol Soc London 142: 515-526

Linnen RL, Williams-Jones AE (1987) Tectonic control of quartz vein orientations at the Trout Lake stockwork molybdenum deposit, southeastern British Columbia: implications for metallogeny in the Kootenay Arc. Econ Geol 82: 1283-1293

Lipman PW (1971) Iron-titanium oxide phenocrysts in compositionally zoned ash-flow sheets from southern Nevada. J Geol 79: 438-456

Lipman PW (1988) Evolution of silicic magma in the upper crust: the mid-Tertiary Latir volcanic field and its cogenetic granitic batholith, northern New Mexico, USA. Transact R Soc Edinburgh, Earth Sci 79: 265-288.

Little W (1960) Inclusions in cassiterite and associated minerals. Econ Geol 55: 485-509

Liu Yingjun, Zhang Jingrong, Sun Chengyuan, Ma Dongsheng, Qiao Enguang, Chen Jun (1984) The geochemical characteristics of trace elements in granitic rocks of South China. Proc Int Symp Geology of Granites and their metallogenetic Relations, Nanjing, Oct 26-30, 1982. Sci Press, Beijing, pp 753-770

London D (1986) Magmatic-hydrothermal transition in the Tanco rare-element pegmatite: evidence from fluid inclusions and phase-equilibrium experiments. Am Mineral 71: 376-395

London D, Hervig RL, Morgan GB VI (1988) Melt-vapor solubilities and elemental partitioning in peraluminous granite-pegmatite systems: experimental results with Macusani glass at 200 MPa. Contrib Mineral Petrol 99: 360-373

Lorenz W, Schirn R (1987) Mylonite, Diaphthorite und epigenetische Zinnmineralisation in der Felsitzone nordwestlich von Freiberg, Erzgebirge. Z Geol Wiss 15: 565-597

Loss RD, Rosman KKR, De Laeter JR (1989) The solar system abundance of tin. Geochim Cosmochim Acta 53: 933-935

Luth WC (1969) The system $NaAlSi_3O_8$-SiO_2 and $KAlSi_3O_8$-SiO_2 to 20 kb and the relationship between H_2O content, p_{H2O}, and p_{total} in granitic magmas. Am J Sci 267-A: 325-341

Luth WC, Jahns RH, Tuttle OF (1964) The granite system at pressures of 4 to 10 kilobars. J Geophys Res 69: 759-773

MacAlister DA (1908) Geological aspect of the lodes of Cornwall. Econ Geol 3: 363-380

MacLeod WN, Turner DC, Wright EP (1971) The geology of the Jos Plateau. Vol 1. General geology. Bull Geol Surv Nigeria 32: 1-110

Magak'yan IG (1968) Ore deposits: tin. Int Geol Rev 10: 108-121

Mahood G, Hildreth W (1983) Large partition coefficients for trace elements in high-silica rhyolites. Geochim Cosmochim Acta 47: 11-30

Manning DAC (1981) The effect of fluorine on liquidus phase relationships in the system Qz-Ab-Or with excess water at 1 kbar. Contrib Mineral Petrol 76: 206-215

Manning DAC, Henderson P (1984) The behaviour of tungsten in granitic melt-vapour systems. Contrib Mineral Petrol 86: 286-293

Manning DAC, Pichavant M (1984) Experimental studies of the role of fluorine and boron in the formation of late-stage granitic rocks and associated mineralisation. Proc 27th Int Geol Congr Moscow (Utrecht, VNU Sci Press) 9: 353-372

Manning DAC, Pichavant M (1988) Volatiles and their bearing on the behaviour of metals in granitic systems. In: Taylor RP, Strong DF (eds) Recent advances in the geology of granite-related mineral deposits. Can Inst Min Metall Spec Vol 39: 13-24

Marsh B (1987) Magmatic processes. Rev Geophysics 25: 1043-1053

Martin JS (1983) An experimental study of the effects of lithium on the granite system. Proc Ussher Soc 5: 417-420

Martinez C (1980) Structure et évolution de la chaîne Hercynienne et de la chaîne Andine dans le nord de la Cordillère des Andes de Bolivie. Trav Docum ORSTOM <Paris> 119: 1-352

Mason DR, McDonald JA (1978) Intrusive rocks and porphyry copper occurrences of the Papua New Guinea-Solomon Islands region: a reconnaissance study. Econ Geol 73: 857-877

Matheis G, Caen-Vachette M (1983) Rb-Sr isotopic study of rare-metal bearing and barren pegmatites in the Pan-African reactivation zone of Nigeria. J Afr Earth Sci 1: 35-40

Maucher A (1965) Die Antimon-Wolfram-Quecksilber-Formation und ihre Beziehungen zu Magmatismus und Geotektonik. Freiberger Forschungsh C 186: 173-188

McBride SL, Robertson RCR, Clark AH, Farrar E (1983) Magmatic and metallogenetic episodes in the northern tin belt, Cordillera Real, Bolivia. Geol Rdsch 72: 685-713

McCallum ME (1985) Experimental evidence for fluidization processes in breccia pipe formation. Econ Geol 80: 1523-1543

McCarthy TS, Cawthorn RG (1980) Changes in initial $^{87}Sr/^{86}Sr$ ratio during protracted fractionation in igneous complexes. J Petrol 21: 245-264

McCarthy TS, Groves DI (1979) The Blue Tier batholith, northeastern Tasmania: a cumulate-like product of fractional crystallization. Contrib Mineral Petrol 71: 193-209

McCarthy TS, Hasty RA (1976) Trace element distribution patterns and their relationship to the crystallization of granitic melts. Geochim Cosmochim Acta 40: 1351-1358

McCulloch MT, Chappell BW (1982) Nd isotopic characteristics of S- and I-type granites. Earth Planet Sci Lett 58: 51-64

McKenzie WF, Helgeson HC (1985) Phase relations among silicates, copper iron sulfides, and aqueous solutions at magmatic temperatures. Econ Geol 80: 1965-1973

McNutt RH, Clark AH (1983) Implications of the initial strontium isotope ratios of Central Andean, Triassic-to-Quaternary, igneous rocks in N Chile, S Peru and NW Bolivia. Eos (Transact Am Geophys Union) 64: p 329

McNutt RH, Crocket JH, Clark AH, Caelles JC, Farrar E, Haynes SJ, Zentilli M (1975) Initial $^{87}Sr/^{86}Sr$ ratios of plutonic and volcanic rocks of the central Andes between latitudes 26° and 29° South. Earth Planet Sci Lett 27: 305-313

MCS (1990) Mineral commodity summaries 1990. Minerals Inform Office (US Bureau Mines/US Geol Surv), Washington, 200 pp

Metallgesellschaft (1965) Statistische Zusammenstellungen über Aluminium, Blei, Kupfer, Zink, Zinn, Kadmium, Magnesium, Nickel, Quecksilber und Silber. Metallgesellschaft, Frankfurt, 264 pp

Metallgesellschaft (1976) Metallstatistik 1965-1975. Metallgesellschaft, Frankfurt, 362 pp

Metallgesellschaft (1989) Metallstatistik 1978-1988. Metallgesellschaft, Frankfurt, 480 pp

Meyer C (1985) Ore metals through geologic history. Science 227: 1421-1428

Michael PJ (1983) Chemical differentiation of the Bishop Tuff and other high silica magma chambers through crystallization processes. Geology 11: 31-34

Michel H, Schneider HJ (1978) Uranvorkommen im Zusammenhang mit den tertiären Vulkaniten des lateinamerikanischen Kordillerenzuges. Erzmetall 31: 1-8

Miller CF, Mittlefehldt DW (1984) Extreme fractionation in felsic magma chambers: a product of liquid-state diffusion or fractional crystallization? Earth Planet Sci Lett 68: 151-158

Miller JF (1988) Granite petrogenesis in the Cordillera Real, Bolivia and crustal evolution in the Central Andes. Unpubl Ph.D. thesis, Milton Keynes, Open Univ, 198 pp

Miller JF, Harris NBW (1989) Evolution of continental crust in the Central Andes; constraints from Nd isotope systematics. Geology 17: 615-617

Mining Journal (1990) Neves-Corvo tin inauguration. Min Journ 314:289

Mitchell AHG (1974) Southwest England granites: magmatism and tin mineralization in a post-collision tectonic setting. Transact Instn Min Metall 83: B95-B97

Mitchell AHG (1977) Tectonic settings for emplacement of southeast Asian tin granites. Bull Geol Soc Malaysia 9: 123-140

Möller P (1989) REE(Y), Nb, and Ta enrichment in pegmatites and carbonatite-alkalic rock complexes. In: Möller P, Cerny P, Saupé F (eds) Lanthanides, tantalum and niobium. Mineralogy, geochemistry, characteristics of primary ore deposits, prospecting, processing and applications. Proceedings of a workshop in Berlin, November 1986. Springer, Berlin Heidelberg New York, pp 103-144

Möller P, Dulski P (1983) Fractionation of Zr and Hf in cassiterite. Chem Geol 40: 1-12

Moon KJ (1988) Sangdong tungsten ore deposits. Guidebook post-Symp Field Conf IGCP 220, Oct 21-23, 1988. Seoul, Korean Organizing Committ 5th Symp IGCP 220, pp 71-87

Moon KJ (1989) Discovery of source rock of Sangdong tungsten mineralization. 28th Int Geol Congr, Washington, Abstr 2: 454

Moore WJ (1978) Chemical characteristics of hydrothermal alteration at Bingham, Utah. Econ Geol 63: 612-621

Moore WJ, Lanphere MA, Obradovich JD (1968) Chronology of intrusion, volcanism, and ore deposition at Bingham, Utah. Econ Geol 63: 612-621

Morgan JW, Anders E (1980) Chemical composition of Earth, Venus, and Mercury. Proc Natl Acad Sci USA 77: 6973-6977

Mosch E, Becker M (1985) Largest cassiterite flotation plant in the world under test operation at Altenberg (GDR). Min Mag 1985: 531-537

Mulligan R (1975) Geology of Canadian tin occurrences. Geol Surv Canada, Econ Geol Report 28: 1-155

Musselwhite DS, DePaolo DJ, McCurry M (1989) The evolution of a silicic magma system: isotopic and chemical evidence from the Woods Mountains volcanic center, eastern California. Contrib Mineral Petrol 101: 19-29

Mysen BO, Virgo D, Seifert FA (1985) Relationships between properties and structure of aluminosilicate melts. Am Mineral 70: 88-105

Mysen BO, Virgo D (1989) Redox equilibria, structure, and properties of Fe-bearing aluminosilicate melts: relationships among temperature, composition, and oxygen fugacity in the system $Na_2O-Al_2O_3-SiO_2-Fe-O$. Am Mineral 74: 58-76

Nakapadungrat S, Chulacharit N, Munthachit Y, Chotigkrai T, Sangsila S (1984a) Geology of Sn-W granites in Takua Pa area, southern Thailand. Paper presented at: Int Symp Geology of Tin Deposits, 27 Oct-8 Nov, 1984, Nanning/China, 34 pp. Abstr in: Report on the Int Symp Geol Tin Deposits. Bandung, ESCAP Regional Mineral Resour Develop Cent, p 96

Nakapadungrat S, Beckinsale RD, Suensilpong S (1984b) Geochronology and geology of Thai granites. Proc Conf Applications of Geology and the National Development, 19-22 Nov 1984. Chulalongkorn Univ, Bangkok, pp 75-93

Neiva AMR (1975) Geochemistry of coexisting aplites and pegmatites and of their minerals from central northern Portugal. Chem Geol 16: 153-177

Neiva AMR (1976) The geochemistry of biotites from granites of northern Portugal with special reference to their tin content. Mineral Mag 40: 453-466

Neiva AMR (1982) Geochemistry of muscovite and some physico-chemical conditions of the formation of some tin-tungsten deposits in Portugal. In: Evans AM (ed) Metallization associated with acid magmatism. Wiley, Chichester, pp 243-259

Neiva AMR (1984) Geochemistry of tin-bearing granitic rocks. Chem Geol 43: 241-256

Nekrasov IY (1984) Tin in magmatic and postmagmatic processes. Nedra, Moscow (in Russian)

Nekrasov IY, Epel'baum MB, Sobolev VP (1981) Study of the model system granite-$SnO(SnO_2)$-fluid: Tin content of quartz-albite melt as a function of oxygen fugacity (fO_2). Dokl Earth Sci Sect 247: 220-223 (translated from Dokl Akad Nauk SSSR 247: 696-699, 1979)

Nekrasov IY, Epel'baum MB, Sobolev VP (1982) Partition of tin between melt and chloride fluid in the granite-SnO-SnO_2-fluid system. Dokl Earth Sci Sect 252: 165-168 (translated from Dokl Akad Nauk SSSR 252: 977-981, 1980)

Neumann H, Mead J, Vitaliano CJ (1954) Trace element variation during fractional crystallization as calculated from the distribution law. Geochim Cosmochim Acta 6: 90-99

Newsom HE, Palme H (1984) The depletion of siderophile elements in the Earth's mantle: new evidence from molybdenum and tungsten. Earth Planet Sci Lett 69: 354-364

Niggli P (1920) Die leichtflüchtigen Bestandteile im Magma. Teubner, Leipzig, 272 pp

Noble DC, Sargent KA, Mehnert HH, Ekren EB, Byers FM Jr (1968) Silent Canyon volcanic center, Nye County, Nevada. Geol Soc Am Mem 110: 65-74

Noble DC, Vogel TA, Peterson PS, Landis GP, Grant NK, Jezek PA, McKee EH (1984) Rare-element enriched, S-type ash-flow tuffs containing phenocrysts of muscovite, andalusite, and sillimanite, southeastern Peru. Geology 12: 35-39

Norton D (1988) Metasomatism and permeability. Am J Sci 288: 604-619

Norton D, Knapp R (1977) Transport phenomena in hydrothermal systems: the nature of porosity. Am J Sci 277: 913-936

Norton D, Knight J (1977) Transport phenomena in hydrothermal systems: cooling plutons. Am J Sci 277: 937-981

Norton D, Taylor JPJr (1979) Quantitative simulation of the hydrothermal systems of crystallizing magmas on the basis of transport theory and oxygen isotope data: an analysis of the Skaergaard Intrusion. J Petrol 20: 421-486

Olade MA (1980) Geochemical characteristics of tin-bearing and tin-barren granites, northern Nigeria. Econ Geol 75: 71-82

Omer-Cooper WRB, Hewitt WV, Wees H van (1974) Exploration for cassiterite-magnetite-sulphide veins on Belitung, Indonesia. Proc 4th World Conf Tin, Kuala Lumpur, pp 97-119

Osberger R (1962) Petaq geologi dari pulau Belitung. Skala 1:100000 (Geological map of Belitung Island. Scale 1:100,000). P.T.Timah, Kelapa Kampit, 1 sheet

Osborn EF (1979) The reaction principle. In: Yoder HS (ed) The evolution of the igneous rocks. Fiftieth anniversary perspectives. Princeton Univ Press, Princeton, pp 133-169

Pälchen W, Rank G, Lange H, Tischendorf G (1987) Regionale Clarkewerte - Möglichkeiten und Grenzen ihrer Anwendung am Beispiel des Erzgebirges (DDR). Chem Erde 47: 1-17

Pan YS (1974) The genesis of the Mexican-type tin deposits in acid volcanics. Ph.D. thesis, Columbia Univ., Univ Microfilms Int, Ann Arbor, 286 pp

Pankhurst RJ, Hole MJ, Brook M (1988) Isotope evidence for the origin of Andean granites. Transact R Soc Edinburgh, Earth Sci 79: 123-133

Patterson DJ, Ohmoto H, Solomon M (1981) Geologic setting and genesis of cassiterite-sulfide mineralization at Renison Bell, western Tasmania. Econ Geol 76: 393-438

Pearce JA, Harris NBW, Tindle AG (1984) Trace element discrimination diagrams for the tectonic interpretation of granitic rocks. J Petrol 25: 956-983

Pei Rongfu, Mao Jingwen (1988) On petro-minerogenetic models of tin-tungsten granite in South China. Extended Abstr, 5th Int Symp Tin-Tungsten Granites in Southeast Asia and the western Pacific, Oct 17-19, 1988. Shimane Univ, Matsue, pp 124-128

Petersen U (1979) Metallogenesis in South America: progress and problems. Episodes 4: 3-11

Petrascheck WE (1933) Die Erzlagerstätten des Schlesischen Gebirges. Archiv Lagerstättenforschung 59: 1-53

Petrova ZI, Legeydo VA (1965) Geochemistry of tin in the magmatic process. Geochem Intern 2: 301-307 (translated from Geokhimiya 1965, 4: 482-489)

Pichavant M (1981) An experimental study of the effect of boron on a water saturated haplogranite at 1 kbar pressure. Geological applications. Contrib Mineral Petrol 76: 430-439

Pichavant M, Valencia Herrera J, Boulmier S, Briqueu L, Joron JL, Juteau M, Marin L, Michard A, Sheppard SMF, Treuil M, Vernet M (1987) The Macusani glasses, SE Peru: evidence of chemical fractionation in peraluminous magmas. In: Mysen BO (ed) Magmatic processes: physicochemical principles. Geochem Soc, Spec Publ 1: 359-373

Pichavant M, Kontak DJ, Valencia Herrera J, Clark AH (1988a) The Miocene-Pliocene Macusani volcanics, SE Peru. I. Mineralogy and magmatic evolution of a two-mica aluminosilicate-bearing ignimbrite suite. Contrib Mineral Petrol 100: 300-324

Pichavant M, Kontak DJ, Briqueu L, Valencia Herrera J, Clark AH (1988b) The Miocene-Pliocene Macusani volcanics, SE Peru. II. Geochemistry and origin of a felsic peraluminous magma. Contrib Mineral Petrol 100: 325-338

Pichler H, Zeil W (1969) Die quartäre "Andesit"-Formation in der Hochkordillere Nord-Chiles. Geol Rdsch 58: 866-903

Pichler H, Zeil W (1972) Paleozoic and Mesozoic ignimbrites of northern Chile. N Jb Miner Abh 116: 196-207

Pitcher WS (1979) The nature, ascent and emplacement of granite magmas. J Geol Soc London 136: 627-662

Pitcher WS (1987) Granites and yet more granites forty years on. Geol Rdsch 76: 51-79

Pitfield PEJ (1987) South-East Asia Granite Project: Report on the geochemistry of the Tin Islands granites of Indonesia. British Geol Surv, Overseas Report MP/87/9/R, pp 1-51

Pitfield PEJ, Cobbing EJ, Mallick DIJ, Clarke MCG, Teoh LH (1987) Granite provinces in the Southeast Asian tin belt. Transact, 4th Circum-Pacific Energy and Mineral Resources Conf, Aug 17-22, 1986, Singapore. Circum-Pacific Counc Energy Mineral Res/AAPG, Tulsa, pp 575-589

Pitfield PEJ, Teoh LH, Cobbing EJ (1990) Two-phase variants and Sn-mineralization in the Main Range Province granites of the Southeast Asian Tin Belt. Bull Geol Soc Malaysia (in press)

Plimer IR (1980) Exhalative Sn and W deposits associated with mafic volcanism as precursors to Sn and W deposits associated with granites. Mineral Deposita 15: 275-289

Plimer IR (1987) Fundamental parameters for the formation of granite-related tin deposits. Geol Rdsch 76: 23-40

Pliny the Elder → Plinius Secundus G (1984) Naturalis Historiae/Naturkunde. Bd 33 (Metallurgie). Artemis, München, 226 pp

Pollard PJ, Taylor RG (1986) Progressive evolution of alteration and tin mineralization: controls by interstitial permeability and fracture-related tapping of magmatic fluid reservoirs in tin granites. Econ Geol 81: 1795-1800

Polya DA (1988) Efficiency of hydrothermal ore formation and the Panasqueira W-Cu-(Ag)-Sn vein deposit. Nature <London> 333: 838-841

Priem HNA, Bon EH, Verdurmen EAT, Bettencourt JS (1989) Rb-Sr chronology of Precambrian crustal evolution in Rondônia (western margin of the Amazon craton), Brazil: J South Am Earth Sci 2: 163-170

Putthapiban P (1984) Geochemistry, geochronology, and tin mineralization of Phuket granites, Phuket, Thailand. Unpubl Ph.D. thesis, La Trobe Univ, Victoria, 414 pp

Putthapiban P, Jackson P, Gray CM (1986) The possible age of the maximum tin/tungsten mineralization at Phuket Island, southern Thailand. Proc Int Symp Correlation and Resource Evaluation of Tin-Tungsten Granites in Southeast Asia and the Western Pacific Region, IGCP Project 220, June 30-July 2, 1986, Canberra. Bureau Mineral Resour Record 1986/10, pp 61-62

Putzer H (1940) Die zinnführende Fahlbandlagerstätte von Giehren am Isergebirge. Z Dtsch Geol Ges 92: 137-158

Putzer H (1942) Die Zinngrube "Reicher Trost" bei Giehren im Isergebirge. Z Dtsch Geol Ges 94: 37-40

Raimbault L (1984) Géologie, pétrographie et géochimie des granites et minéralisations associées de la région de Meymac (Haute Corrèze, France). Thèse, École Nat Sup Mines Saint-Étienne, Saint-Étienne, 482 pp

Rajah SS (1979) The Kinta tinfield, Malaysia. Bull Geol Soc Malaysia 11: 111-136

Ranchin G (1970) La géochimie de l'uranium et la différenciation granitique dans la province uranifère du Nord-Limousin. Thèse, Univ Nancy, Nancy, 394 pp

Rapela CW, Shaw DM (1979) Trace and major element models of granitoid genesis in the Pampean Ranges, Argentina. Geochim Cosmochim Acta 43: 1117-1129

Rapela CW, Heaman LM, McNutt RH (1982) Rb-Sr geochronology of granitoid rocks from the Pampean Ranges, Argentina. J Geol 90: 574-582

Rayleigh JWS (1896) Theoretical considerations respecting the separation of gases by diffusion and similar processes. Philos Mag 42: 77-107

Redwood SD (1986) Epithermal precious and base metal mineralisation and related magmatism of the northern Altiplano, Bolivia. Unpubl Ph.D. thesis, Univ Aberdeen, Aberdeen, 229 pp

Reed BL, Menzie WD, McDermott M, Root DH, Scott W, Drew LJ (1989) Undiscovered lode tin resources of the Seward Peninsula, Alaska. Econ Geol 84, 1936-1947

Reyer E (1881) Zinn: Eine geologisch-montanistisch-historische Monografie. Reimer, Berlin, 248 pp

Reynolds DL (1954) Fluidization as a geological process, and its bearing on the problem of intrusive granites. Am J Sci 252: 577-614

Rice A (1981) Convective fractionation: a mechanism to provide cryptic zoning (macrosegregation), layering, cres-cumulates, banded tuffs and explosive volcanism in igneous processes. J Geophys Res 86 (B1): 405-417

Richter H (1987) Ergebnisse geologischer Forschungs- und Erkundungsarbeiten und Aufgabenstellung für ihre weitere Entwicklung nach dem XI. Parteitag. Z Angew Geol 33: 57-64

Richter P (1984) Wolfram in Graniten Ostbayerns. Versuch einer metallogenetischen Gliederung. Geol Jb < Hannover > D 63: 3-22

Richter P, Stettner G (1979) Geochemische und petrographische Untersuchungen der Fichtelgebirgsgranite. Geologica Bavarica 78: 1-144

Richter P, Stettner G (1987) Die Granite des Steinwaldes (Nordost-Bayern) - ihre petrographische und geochemische Differenzierung. Geol Jb < Hannover > D 86: 3-31

Ringwood AE (1979) Origin of the Earth and Moon. Springer, New York Heidelberg Berlin, 295 pp

Roedder E (1984) Fluid inclusions. Rev Mineral 12: 1-646

Rösler HJ, Lange H (1976) Geochemische Tabellen, 2nd ed. Enke, Stuttgart, 674 pp

Rößler B (1700) Speculum metallurgiae politissimum. Oder: Hell-polierter Berg-Bau-Spiegel. Johann Jacob Winckler, Dresden, 168 pp

Rose AW, Burt DM (1979) Hydrothermal alteration. In: Barnes H (ed) Geochemistry of hydrothermal ore deposits, 2nd ed. Wiley, New York, pp 173-235

Routhier P (1967) Eassai critique sur les méthodes de la géologie. De l'objet à la genèse. Masson, Paris, 204 pp

Routhier P (1980) Où sont les métaux pour l'avenir? Les provinces métalliques. Essai de métallogénie globale. Mém BRGM 105: 1-410

Routhier P (1983) Where are the metals for the future? The metal provinces. An essay of global metallogeny. BRGM, Orléans, 400 pp

Rub MG (1968) Die Besonderheiten der stofflichen Zusammensetzung und Genesis zinnführender magmatischer Komplexe und der Charakter ihrer Vererzung. Z Angew Geol 14: 193-204

Ruiz J (1988) Petrology, distribution and origin of rhyolites associated with tin mineralization in the Sierra Madre Occidental, Mexico. In: Taylor RP, Strong DF (eds) Recent advances in the geology of granite-related mineral deposits. Can Inst Min Metall, Spec Vol 39: 322-330

Rutherford MJ (1969) An experimental determination of iron biotite-alkali feldspar equilibria. J Petrol 10: 381-408

Ryabchikov ID, Durasova NA, Barsukov VL, Yefimov AS (1978) The redox potential as a factor governing ore production by an acid magma. Geochem Intern 1978, 3: 121-123 (translated from Geokhimiya 1978, 6: 832-834)

Saavedra J, Toselli AJ, Rossi de Toselli JN, Rapela CW (1987) Role of tectonism and fractional crystallization in the origin of lower Paleozoic epidote-bearing granitoids, northwestern Argentina. Geology 15: 709-713

Sack RO, Carmichael ISE, Rivers M, Ghiorso MS (1980) Ferric-ferrous equilibria in natural silicate liquids at 1 bar. Contrib Mineral Petrol 75: 369-376

Sandberger F (1885) Untersuchungen über Erzgänge. Zweites Heft. CW Kreidel's, Wiesbaden, pp 159-431

Scheepers R, Schoch AE (1988) Geology and geochemistry of the Klipberg alkali feldspar granite and associated hydrothermally altered rocks in the Darling batholith, southwestern Cape Province. S Afr J Geol 91: 212-225

Schermerhorn LJG (1987) The Hercynian gabbro-tonalite-granite-leucogranite suite of Iberia: geochemistry and fractionation. Geol Rdsch 76: 137-145

Schneider A (1985) Eruptive processes, mineralization and isotopic evolution of the Los Frailes-Karikari region/Bolivia. Ph.D. thesis, Univ London, London, 280 pp

Schneider HJ, Lehmann B (1977) Contribution to a new genetical concept on the Bolivian tin province. In: Klemm DD, Schneider HJ (eds) Time- and strata-bound ore deposits. Springer, Berlin Heidelberg, pp 153-168

Schoch AE, Scheepers R (1990) The distribution of uranium and thorium in the Cape Columbine granite from the southwestern Cape Province, South Africa. Ore Geol Rev 5: 223-246

Schröcke H (1954) Zur Paragenese erzgebirgischer Zinnerzlagerstätten. N Jb Miner Abh 87: 33-109

Schröcke H (1986) Die Entstehung der endogenen Erzlagerstätten. De Gruyter, Berlin New York, 878 pp

Schröcke H, Weiner KL (1981) Mineralogie. Ein Lehrbuch auf systematischer Grundlage. De Gruyter, Berlin New York, 952 pp

Schütze H, Stiehl G, Haberlandt R, Gerstenberger H, Wand U, Mühle K, Hermichen WD, Geisler M, Strauch G, Böttger T, Christoph G, Hasse G, Habedank M, Nitzsche HM (1984) Isotopen- und element-geochemische sowie radiogeochronologische Untersuchungen an der Zinnlagerstätte Ehrenfriedersdorf (Erweiterte Zusammen-fassung). ZFI-Mitt <Leipzig> 90: 5-8

Schuiling RD (1967) Tin belts on the continents around the Atlantic Ocean. Econ Geol 62: 540-550

Schwartz M, Askury AK (1988) Geology and geochemistry of the Bujang Melaka tin granite, Kinta Valley, Malaysia. Abstr., 5th Int Symp

Tin/Tungsten Granites in Southeast Asia and the western Pacific. Shimane Univ, Matsue, pp 164-167

Schwartz M, Askury AK (1989) Geologic, geochemical, and fluid inclusion studies of the tin granites of the Bujang Melaka Pluton, Kinta Valley, Malaysia. Econ Geol 84: 751-779

Schwartz M, Surjono (1986) Geology and geochemistry of the Nam Salu horizon, Belitung Island, Indonesia. Proc Int Symp Correlation and Resource Evaluation of Tin/Tungsten Granites in Southeast Asia and the Western Pacific Region, 30 June-2 July, 1986. Bureau Mines Mineral Res Record 1986/10, Canberra, pp 67-68

Schwartz MO, Surjono (1988) The Sn-W greisen deposit at Tikus, Belitung, Indonesia. Abstr, 5th Int Symp Tin/Tungsten Granites in Southeast Asia and the Western Pacific. Shimane Univ, Matsue, pp 171-174

Sclater JG, Parsons B, Jaupart C (1981) Oceans and continents: similarities and differences in the mechanism of heat loss. J Geophys Res 86: 11535-11552

Seltmann R (1990) Bewertung von Suchergebnissen in einem Lagerstätten-bezirk, dargestellt am Beispiel von Zinnerzlagerstätten im Ost-erzgebirge. Veröffentl Zentralinstitut Physik der Erde <Potsdam>, in press [Ph.D. thesis, Bergakademie Freiberg, 1986]

Seltmann R, Baumann L, Legler C (1985) Zur paragenetischen und gefügemässigen Stellung des Magnetits in den Kalksilikatfelsen von Ehrenfriedersdorf. Z Angew Geol 31: 245-250.

Seltmann R, Hösel G, Kühne R, Tischendorf G (1989) Sn-W depositions. In: Tischendorf G (ed) Silicic magmatism and metallogenesis of the Erzgebirge. Veröffentl Zentralinstitut Physik der Erde <Potsdam> 107: 121-148

Seltmann R, Bankwitz P, Bankwitz E, Wetzel HU, Thomas R, Kühne R, Seidel B, Felix M, Märtens S (1990) Crustal structure and intrusion mechanism: the cataclasite-subvolcanic complexes in the Erzgebirge. Preprint to poster, Symp Deformation Processes and the Structure of the Lithosphere, May 3-10, 1990. Acad Sci GDR, Central Inst Physics Earth, Potsdam, 12 pp.

Shannon RD (1976) Revised effective ionic radii and systematic studies of interatomic distances in halides and chalcogenides. Acta Cryst A 32: 751-767

Shepherd TJ, Miller MF, Scrivener RC, Darbyshire DPF (1985) Hydrothermal fluid evolution in relation to mineralization in southwest England with special reference to the Dartmoor-Bodmin area. In: High heat production (HHP) granites, hydrothermal circulation and ore genesis. Instn Mining Metallurgy, London, pp 345-364

Shibata K, Ishihara S, Ulriksen E (1984) Rb-Sr ages and initial $^{87}Sr/^{86}Sr$ ratios of Late Paleozoic granitic rocks from northern Chile. Bull Geol Surv Japan 35: 537-545

Shinohara H, Iiyama JT, Matsuo S (1989) Partition of chlorine compounds between silicate melt and hydrothermal solutions: I. Partition of NaCl-KCl. Geochim Cosmochim Acta 53: 2617-2630

Sillitoe RH (1974) Tin mineralisation above mantle hot spots. Nature <London> 248: 497-499

Sillitoe RH (1976) Andean mineralization: a model for the metallogeny of convergent plate margins. Geol Assoc Canada, Spec Pap 14: 59-100

Sillitoe RH, Halls C, Grant JN (1975) Porphyry tin deposits in Bolivia. Econ Geol 70: 913-927

Sítek J, Stemprok M, Voldán J, Cicáková O (1981) Mössbauer spectra of tin-containing granite glasses. (In Czech, with English summary). Silikáty 25: 243-249

Smith FG (1948) Transport and deposition of the non-sulphide vein minerals. III. Phase relations at the pegmatitic stage. Econ Geol 43: 535-546

Smith TE (1979) The geochemistry and origin of the Devonian granitic rocks of southwest Nova Scotia. Geol Soc Am Bull 90: 850-885

Smith TE, Turek A (1976) Tin-bearing potential of some Devonian granitic rocks in S.W. Nova Scotia. Mineral Deposita 11: 234-245

Smith TE, Miller PM, Huang CH (1982) Solidification and crystallization of a stanniferous granitoid pluton, Nova Scotia, Canada. In: Evans AM (ed) Metallization associated with acid magmatism. Wiley, Chichester, pp 301-320

Snee LW, Sutter JF, Kelly WC (1988) Thermochronology of economic mineral deposits: dating the stages of mineralization at Panasqueira, Portugal, by high-precision $^{40}Ar/^{39}Ar$ age spectrum techniques on muscovite. Econ Geol 83: 335-354

Sparks RSJ, Huppert HE, Turner FRS (1984) The fluid dynamics of evolving magma chambers. Phil Transact R Soc London A 310: 511-534

Speczik S, Wiszniewska J (1984) Some comments about stratiform tin deposits in the Stara Kamienica chain (southwestern Poland). Mineral Deposita 19: 171-175

Speer JA, Naeem A, Almohandis AA (1989) Small-scale variations and subtle zoning in granitoid plutons: the Liberty Hill pluton, South Carolina, U.S.A. Chem Geol 75: 153-181

Stemprok M (1967) Genetische Probleme der Zinn-Wolfram-Vererzung im Erzgebirge. Mineral Deposita 2: 102-118

Stemprok M (1986) Petrology and geochemistry of the Czechoslovak part of the Krusné hory Mts. granite pluton. Sbornik Geol Ved, Lozisk Geol Mineral 27:111-156

Stemprok M (1987) Zinn- und Wolframmineralisation im Böhmischen Massif. Freiberger Forschungshefte C 425: 93-106

Stemprok M (1989) The behavior of tin, tungsten and molybdenum in felsic magmas. Abstr, 28th Int Geol Congr, Washington, 3: 174-175

Stemprok M (1990) Solubility of tin, tungsten and molybdenum oxides in felsic magmas. Mineral Deposita 25 (in press)

Stiehl G (1985) Komplette isotopengeochemische, geochemische und geochronologische Untersuchungen am Granit von Ehrenfrieders-dorf. Z Geol Wiss 13: 585-592

Stoll WC (1964) Metallogenetic belts, centers, and epochs in Argentina and Chile. Econ Geol 59: 126-135

Stone M (1982) The behaviour of tin and some other trace elements during granite differentiation, west Cornwall, England. In: Evans AM (ed) Metallization associated with acid magmatism. Wiley, Chichester, pp 339-355

Stone M, Exley CS (1986) High heat production granites of southwest England and their associated mineralization: a review. Transact Instn Min Metall 95: B25-B36

Stone M, Exley CS, George MC (1988) Compositions of trioctahedral micas in the Cornubian batholith. Mineral Mag 52: 175-192

Strong DF (1988) A review and model for granite-related mineral deposits. In: Taylor RP, Strong DF (eds) Recent advances in the geology of granite-related mineral deposits. Can Inst Min Metall, Spec Vol 39: 424-445

Sun SS, Eadington PJ (1987) Oxygen isotope evidence for the mixing of magmatic and meteoric waters during tin mineralization in the Mole Granite, New South Wales, Australia. Econ Geol 82: 43-52

Swanson SE (1977) Relation of nucleation and crystal growth rate to the development of granitic textures. Am Mineral 62: 966-978

Swanson SE, Bond JF, Newberry RJ (1988) Petrogenesis of the Ear Mountain Tin Granite, Seward Peninsula, Alaska. Econ Geol 83: 46-61

Szalamacha M, Szalamacha J (1974) Geologiczna i petrograficzna charakterystyka lupkow zmineralizowanych kasyterytem na przykladzie kamieniolomu w krobicy (with English summary). Biul Inst Geol 279 (23): 55-90

Tacker RC, Candela PA (1987) Partitioning of molybdenum between magnetite and melt: a preliminary experimental study of partitioning of ore metals between silicic magmas and crystalline phases. Econ Geol 82: 1827-1838

Takahashi M, Aramaki S, Ishihara S (1980) Magnetite-series/ilmenite-series vs. I-type/S-type granitoids. Min Geol, Spec Issue 8: 13-28

Tammemagi HY, Smith NL (1975) A radiogeologic study of the granites of SW England. J Geol Soc London 131: 415-427

Tanelli G, Lattanzi P (1985) The cassiterite-polymetallic sulfide deposits of Dachang (Guangxi, People's Republic of China). Mineral Deposita 20: 102-106

Taylor HP Jr (1977) Water/rock interactions and the origin of H_2O in granitic batholiths. J Geol Soc London 133: 509-558

Taylor HP Jr (1979) Oxygen and hydrogen isotope relationships in hydrothermal mineral deposits. In: Barnes HL (ed) Geochemistry of hydrothermal ore deposits, 2nd edn. Wiley, New York, pp 236-277

Taylor JR, Wall VJ (1984) The mobilisation of tin from granitoid magmas. 27th Int Geol Congr Moscow, Abstr 4(9): p 474

Taylor RG (1979) Geology of tin deposits. Elsevier, Amsterdam, 543 pp

Taylor RG, Pollard PJ, Tate NM (1985) Resource evaluation of primary tin potential of eastern peninsular Malaysia. A district analysis. UNDP/ESCAP Consultancy Mission Report, Southeast Asia Tin Research Developm Cent, Ipoh, Malaysia 12: 1-118

Taylor SR, McLennan SM (1983) Geochemical application of spark-source mass spectrometry. IV. The crustal abundance of tin. Chem Geol 39: 273-280

Taylor SR, McLennan SM (1985) The continental crust: its composition and evolution. Blackwell, Oxford, 312 pp

Teh GH (1981) The Tekka tin deposit, Perak, Peninsular Malaysia. Bull Geol Soc Malaysia 14: 101-118

Terashima S, Ishihara S (1982) Tin abundance of some geosynclinal shales from Japan. J Japan Assoc Min Petr Econ Geol 77: 1-6

Thamm N (1943) Die Zinnerzvorkommen und der Zinnbergbau Afrikas. Mittlg Forschungsstelle Kolonial Bergbau Bergakad Freiberg 3: 1-117

Thomas R (1982) Ergebnisse der thermobarogeochemischen Unter-suchungen an Flüssigkeitseinschlüssen in Mineralen der postmagmatischen Zinn-Wolfram-Mineralisation des Erzgebirges. Freiberger Forschungsh C 370: 1-85

Thomas R, Leeder O (1986) Zur physikochemischen Evolution mineral-bildender Lösungen an Hand thermobarometrischer Unter-suchungen. Z Geol Wiss 14: 51-60

Thomas R, Tischendorf G (1987) Evolution of Variscan magmatic-metallogenetic processes in the Erzgebirge according to thermo-metric investigations. Z Geol Wiss 15: 25-42

Thompson AB, Connolly JAD (1990) Crustal anatexis in orogenic belts: modeling constraints on fluid behavior and melt generation. Eos, Transact Am Geophys Union 71: 650.

Thormann CH, Drew LJ (1988) A report on some of the largest tin deposits in Brazil. US Geol Surv, Admin Rep (Brazil Tin Report), pp 1-19

Thorn PG (1988) Fluid inclusion and stable isotope studies at the Chicote tungsten deposit, Bolivia. Econ Geol 83: 62-68

Tilton GR, Pollak RJ, Clark AH, Robertson RCR (1981) Isotopic composition of Pb in Central Andean ore deposits. Geol Soc Am Mem 154: 791-816

Tin International (1989) Go ahead for Portuguese Mine. Tin Int, January 1989, p 4

Tischendorf G (1970) Zur geochemischen Spezialisierung der Granite des Westerzgebirgischen Teilplutons. Geologie 19: 25-40

Tischendorf G (1988) On the genesis of tin deposits related to granites: the example Erzgebirge. Z Geol Wiss Berlin 16: 407-420

Tischendorf G (1989) Silicic magmatism and metallogenesis of the Erzgebirge (compiled by G. Tischendorf). Veröffentlich Zentralinst Physik Erde < Potsdam > 107: 1-316

Tischendorf G, Schust F, Lange H (1978) Relation between granites and tin deposits in the Erzgebirge, GDR. In: Stemprok M, Burnol L, Tischendorf G (eds) Metallization associated with acid magmatism, Vol. 3. Czech Geol Surv, Praha, pp 123-137

Tischendorf G, Geisler M, Gerstenberger H, Budzinski H, Vogler P (1987) Geochemistry of Variscan granites of the Westerzgebirge-Vogtland region - an example of tin deposit-generating granites. Chem Erde 46: 213-235

Tistl M (1985) Die Goldlagerstätten der nördlichen Cordillera Real/Bolivien und ihr geologischer Rahmen. Berliner Geowiss Abh A 65: 1-93

Trumbull RB, Morteani G (1986) The Mhlosheni Pluton, an Archean A-type granite in Swaziland and its relation to tin mineralization. Abstr, 76. Jahrestagung Geol Vereinigung, 26.2.-1.3.1986, Giessen, p 78

Turekian KK, Wedepohl KH (1961) Distribution of the elements in some major units of the Earth's crust. Bull Geol Soc Am 72: 175-191

Tuttle OF, Bowen NL (1958) Origin of granite in the light of experimental studies in the system $NaAlSi_3O_8$-$KAlSi_3O_8$-SiO_2-H_2O. Geol Soc Am Mem 74: 1-153

Urabe T (1985) Aluminous granite as a source of hydrothermal ore deposits: an experimental study. Econ Geol 80: 148-157

Veizer J (1988) Continental growth: comments on "The Archean-Proterozoic transition: evidence from Guyana and Montana" by Gibbs AK, Mont-gomery CW, O'Day PA, Erslev EA. Geochim Cosmochim Acta 52: 789-792

Veizer J, Jansen SL (1979) Basement and sedimentary recycling and continental evolution. J Geol 87: 341-370

Veizer J, Jansen SL (1985) Basement and sedimentary recycling 2. Time dimension to global tectonics. J Geol 93: 625-643

Veizer J, Laznicka P, Jansen SL (1989) Mineralization through geologic time: recycling perspective. Am J Sci 289: 484-524

Wänke H (1981) Constitution of terrestrial planets. Phil Trans R Soc London A 303: 287-302

Wänke H, Dreibus G (1988) Chemical composition and accretion history of terrestrial planets. Phil Trans R Soc London A 325: 545-557

Wänke H, Dreibus G, Jagoutz E (1984) Mantle chemistry and accretion history of the Earth. In: Kröner A, Hanson GN, Goodwin AM (eds) Archaean geochemistry. Springer, Berlin Heidelberg New York Tokyo, pp 1-24

Watmuff G (1978) Geology and alteration-mineralization zoning in the central portion of the Yandera porphyry copper prospect, Papua New Guinea. Econ Geol 73: 829-856

Watson EB, Harrison TM (1983) Zircon saturation revisited: temperature and composition effects in a variety of crustal magma types. Earth Planet Sci Lett 64: 295-304

Watznauer A (1954) Die erzgebirgischen Granitintrusionen. Geologie 3: 688-706

Webster JD (1990) Partitioning of F between H_2O and CO_2 fluids and topaz rhyolite melt. Implications for mineralizing magmatic-hydrothermal fluids in F-rich granitic systems. Contrib Mineral Petrol 104: 424-438

Weinhold G (1977) Zur prävaristischen Vererzung im Erzgebirgskristallin aus der Sicht seiner lithofaziellen und geotektonisch-magmatischen Entwicklung während der assyntisch-kaledonischen Ära. Freiberger Forschungsh C 320: 1-53

Werner AG (1791) Neue Theorie von der Entstehung der Gänge, mit Anwendung auf den Bergbau besonders den freibergischen. Gerlachische Buchdrukkerei, Freiberg, 256 pp

Whalen JB (1980) Geology and geochemistry of the molybdenite showings of the Ackley City batholith, southeast Newfoundland. Can J Earth Sci 17: 1246-1258

Whalen JB, Britten RM, McDougall I (1982) Geochronology and geochemistry of the Frieda River prospect area, Papua New Guinea. Econ Geol 77: 592-616

White AJR, Chappell BW (1977) Ultrametamorphism and granitoid genesis. Tectonophysics 43: 7-22

White AJR, Chappell BW (1983) Granitoid types and their distribution in the Lachlan Fold Belt, southeastern Australia. Geol Soc Am Mem 159: 21-34

White WH, Bookstrom AA, Kamilli RJ, Ganster MW, Smith RP, Ranta DE, Steininger RC (1981) Character and origin of Climax-type molybdenum deposits. Econ Geol, 75th Anniv Vol, pp 270-316

Whitney JA (1975) Vapor generation in a quartz monzonite magma: a synthetic model with application to porphyry copper deposits. Econ Geol 70: 346-358

Willis-Richards J, Jackson NJ (1989) Evolution of the Cornubian ore field, southwest England: Part I. Batholith modeling and ore distribution. Econ Geol 84: 1078-1100

Wilson GA, Eugster HP (1984) Cassiterite solubility and metal-chloride speciation in supercritical solutions. Geol Soc Am Ann Meeting, Abstr with Programs 16: 696

Winkelmann L (1983) Geologie und Lagerstätten im Bereich Palca (Mururata) und die Geochemie der Silursedimentite in der Cordillera La Paz/Bolivien. Berliner Geowiss Abh A 51: 1-110

Wohletz KH, Sheridan MF (1979) A model of pyroclastic surge. Geol Soc Am Spec Pap 180: 177-194

Wolf M, Sanchez J (1976) Zur Stellung des Wismuts in den Erzparagenesen einiger bolivianischer Lagerstätten. Freiberger Forschungsh C 315: 53-90

Wones DR (1989) Significance of the assemblage titanite + magnetite + quartz in granitic rocks. Am Mineral 74: 744-749

Wood SA, Vlassopoulos D (1989) Experimental determination of the hydrothermal solubility and speciation of tungsten at 500 C and 1 kbar. Geochim Cosmochim Acta 53: 303-312

Wood SA, Crerar DA, Borcsik MP (1987) Solubility of the assemblage pyrite-pyrrhotite-magnetite-sphalerite-galena-gold-sibnite-bismuthinite-argentite-molybdenite in H_2O-NaCl-CO_2 solutions from 200 to 350 C. Econ Geol 82: 1864-1887

Wyllie PJ, Huang WL, Stern CR, Maaloe S (1976) Granitic magmas: possible and impossible sources, water contents, and crystallization sequences. Can J Earth Sci 13: 1007-1019

Yang Shiyi, Liu Houqun, Zhang Xiulan, Chen Changjiang (1984) The metallogenic characteristics of the Xiling porphyry tin (copper) polymetallic deposit in eastern Guangdong province with an approach to the source of its rock-forming and ore-forming substances. Int Symp Geology tin deposits (Nanning and Dachang, 27 Oct-8 Nov, 1984). Chin Acad Geol Sci/ESCAP Regional Min Res Develop Progr, Bandung, p 109

Ypma PJM, Simons JH (1969) Genetical aspects of the tin mineralization in Durango, Mexico. In: Fox W (ed) A second technical conference on tin, Vol. 1. Int Tin Counc/Dept Mineral Res Thailand, Bangkok, pp 179-192

Zimmermann CF (1746) Ober-Sächsische Berg-Academie, in welcher die Bergwercks-Wissenschaften nach ihren Grund-Wahrheiten untersuchet, und nach ihrem Zusammenhange entworffen werden. Friedrich Hekel, Dresden Leipzig, 288 pp

Zimmermann C (1808) Darstellungen aus der Mineralogie, Mathematik, Physik und Bergwerkskunde. Mohr und Zimmer, Heidelberg, 310 pp

Zippe FXM (1857) Geschichte der Metalle. Braumüller, Wien, 364 pp

Locality Index

Subject Index

Lecture Notes in Earth Sciences